BUDGET FLYING

Gulfstream American Corporation

by
Dan Ramsey

McGraw-Hill Book Company

New York   Johannesburg   Panama
St. Louis   London   Paris
San Francisco   Madrid   São Paulo
Auckland   Mexico   Singapore
Bogotá   Montreal   Sydney
Hamburg   New Delhi   Tokyo

BUDGET FLYING

How to Earn Your
Private Pilot
License and
Enjoy Flying
Economically

McGraw-Hill Series in Aviation

David B. Thurston Consulting Editor

Ramsey *Budget Flying* (1981)
Sacchi *Ocean Flying* (1979)
Smith *Aircraft Piston Engines* (1981)
Thomas *Personal Aircraft Maintenance* (1981)
Thurston *Design for Flying* (1978)
Thurston *Design for Safety* (1980)
Thurston *Homebuilt Aircraft* (1981)

Library of Congress Cataloging in Publication Data

Ramsey, Dan, date.
  Budget flying.

  (McGraw-Hill series in aviation)
  Includes index.
  1. Private flying—United States. I. Title.
II. Series.
TL721.4.R35   629.132'5217   80-18619
ISBN 0-07-051202-7

1 2 3 4 5 6 7 8 9 0   DODO   8 9 8 7 6 5 4 3 2 1 0

The editors for this book were Jeremy Robinson and Olive Collen, the
designer was Mark E. Safran, and the production supervisor was
Paul A. Malchow. It was set in Optima by The Fuller Organization.

Printed and bound by R. R. Donnelley & Sons Company.

To our chosen children—
Byron Hyun Mo and Brendon Yun Mo

CONTENTS

PREFACE

"Mooney forty-four Victor departing on runway two-six."

My friend Dick Williamson snapped the mike back into its bracket on the control panel and pushed in the throttle. We began rolling down the asphalt ribbon. The small plane bounced lightly in its effort to break the bonds of earth and climb into the sky. The speed built: 45 . . . 50 . . . 55 . . . 60 . . . 65 . . . , and then the 2500-pound plane obeyed the laws of physics and swiftly lifted off the runway.

At that moment I knew I wanted to fly.

The problem I saw was that you had to be rich to be a pilot. Planes cost $25,000 or more and required maintenance by $30-per-hour licensed mechanics. Instruction for a private pilot license could cost as much as $2000.

My dreams of flying crashed!

Then I turned to Dick and other pilots I knew to find out how someone with an average income can earn a private pilot license and enjoy flying economically. Dick works for the railroad. Other pilot-friends are small-business owners, housewives, salespeople, and factory workers. How can they afford to fly? How did they earn their wings? How can they pay the high price of going up?

This book, *Budget Flying*, gives practical answers to those and dozens of other questions I posed to hundreds of budget fliers. The fliers talked about pleasure flying and partnerships, instrument ratings and instructors, fuel efficiency and flying clubs. They came up with the right answers—the answers that work. Their hope and mine is that you too can discover the thrill of flight through budget flying.

Like the screen credits at the beginning of a movie, acknowledgment should be given to those who have made special contributions to the development of a book. I would like to acknowledge the help of the following

xi

organizations: Aircraft Owners and Pilots Association, *The Aviation Consumer*, Beech Aircraft Corporation, Bellanca Aircraft Corporation, Cessna Aircraft Company, Collins Division of Rockwell International, Embry-Riddle Aeronautical University, Ercoupe Owners Club, Federal Aviation Administration, General Aviation Electronics, General Aviation Manufacturers Association, Gulfstream American Corporation, Maule Aircraft Corporation, Mill Plain Flying Service, Piper Aircraft Corporation, Pitts Aerobatics, RCA Avionics Systems, Rallye Aircraft Corporation, Robertson Aircraft Corporation, The Soaring Society of America, Taylorcraft Aviation Corporation, Univair Aircraft Corporation, and Wag-Aero Incorporated.

Individually, I offer a special thanks to Dick Williamson, who first took me flying, Mark Johnson and Jim Bowen, who helped me with my wings, David Thurston for his practical advice in developing this book, Jeremy Robinson for his faith and blue pencil, Rick Balkin for his persistence, and Judy for herself.

Vancouver, Washington   **Dan Ramsey**

1 FLYING FOR THE NOT-SO-RICH

Flying began as a hobby for two men in their thirties who had more imagination than they had money. They spent long days and months working toward the moment they could fly. They gathered parts where they could. They tried and failed and learned. Then, on a windy day in 1903, the Wright brothers achieved an ages-old dream of mankind—flight.

From that relatively simple start over 75 years ago we worked toward making flying more complex. Now we have pressurized cockpits, encoding altimeters, glide slope receivers, transponders, and reams of regulations. For each, we have had to pay more and more money.

But something remarkable also happened. A few pioneering men and women began a quiet and individual move back to basic flying at lower costs. They purchased simpler airplanes, learned how to do some of their own repairs, and even found ways to make a few extra dollars to support their hobby. They became budget fliers.

Those economy-minded fliers are a renewed breed. They are motivated by the same dream as the Wright brothers and they face the same problem: how to support their need to fly safely on a limited budget. They are not the rich or the super-rich. They are the not-so-rich. They aren't doctors, bank officers, or professional people. They are factory workers, housewives, salesmen, train engineers, middle-management executives, and truck drivers.

Budget fliers are transported by the sensation of flight. They discover themselves in the sky and refuse to land—even when faced with these current statistics:

The price of the most popular new planes begins at $30,000 and climbs to over $100,000.
Even used airplanes, often older than their pilots, cost more than new cars.

Rental fees for a new plane begin at $20 per hour.

Instructors charge $10 to $18 per hour to teach flying skills.

The private pilot license usually costs from $1200 to $2000 to earn.

Aircraft fuel is even more expensive than automobile fuel.

FAA regulations limit do-it-yourself repairs, and most repairs must be made by licensed mechanics.

The budget fliers are not deterred. To enjoy their hobby they become inventive consumers. They discover sources of safe, low-cost flight instruction. They choose their planes more carefully. They develop cost-cutting partnerships with other economy-minded fliers. They rent older planes. They build planes from kits in their spare time. They learn how to reduce operating costs with consumer techniques. They do as much of their own maintenance as they legally can. They discover the many ways to supplement their flying budgets with extra income.

That's what budget flying is all about. It's discovering and enjoying the thrill of flight for a limited amount of money. Flying isn't cheap—and it should never be funded when it isn't safe—but there are many practical ways in which the smart consumer can earn a private pilot license and enjoy flying economically.

Budget flying is growing. The Aircraft Owners and Pilots Association reports that about 40 percent of its nearly quarter-million members are nonprofessionals and that a third of its members earn less than $25,000 per year.

Anchor Your Head in the Clouds

First the critical question: Do you *really* want, in your spare time, to thrust a 1200-pound collection of nuts and bolts into the sky with you aboard?

Actually, there are three parts to the question:

1. Why do I want to fly?
2. How much is it going to cost me?
3. How do I get started?

First you have to find your own personal reasons for flying. You'll want to do a little soul-searching. You'll want to ask other people why they decided to fly and what they get from the experience.

Next you'll do some wallet-searching. You'll sit down with pencil and paper to find out what flying is going to cost you. You'll discover what "average" pilots spend their money on and how they keep their costs down. You'll also look at a few options to outright ownership of your wings.

Then you'll make your decision and start moving down the runway. You'll set up your own flying budget. You'll draw your own map of where you want to go in flying and how you want to get there. You'll become an authentic budget flier.

The decision is yours. Our technological and economic system has made a 4000-year-old fantasy realizable to nearly anyone who wants to reach out for it. You can discover the thrill of flying—and on a budget.

Today's Flier

General aviation is big business. In the United States, more than 1.2 million pilots and technical ground support people hold active certificates. They fly nearly 200,000 aircraft, *98 percent* of which are general-aviation (nonairline) planes. They have nearly 14,000 public and private airfields to choose from. What is just as surprising to the nonflier is that, of the more than 4½ billion miles flown each year, 80 percent are flown by general-aviation and only 20 percent by airline planes.

Now let's break the 80 percent share down further. About 35 percent of the 4½ billion plus yearly miles are flown by business aircraft; 25 percent are flown for personal reasons; and the remaining 20 percent are used for instruction. And although the percentages may vary slightly from year to year, they continue to say the same thing: general aviation is the quiet ruler of the skies.

Why do so many businesses, devoted as they are to the principle of profit, purchase or lease their own aircraft to move their personnel from one location to another? Actually there are many reasons, but the biggest one is efficiency. Since most business aircraft can fly in a straight line at 150 miles or more per hour with fuel economy that approaches the automobile's, there is the time efficiency factor. A Portland, Oregon, executive can fly to Los Angeles, conduct a full day's business, and fly home in the time it would take to drive the 1000 miles.

In many cases, flying business aircraft is more time-efficient than using scheduled airlines even though the commercial jets travel at many times the speed of the single- or twin-engine propeller plane. The time is saved by not waiting for a scheduled flight, waiting to board, waiting to land at a busy metropolitan airport, and then waiting for luggage. Also, because only about four percent of the nation's airfields are open to commercial jet liners, many passengers must be ferried across town to reach their destinations. A small business plane can often land within a couple of miles of where the pilot wants to be.

Money is no small consideration to the business flier. Despite today's higher cost of planes and fuel, many businesses can own or lease their own aircraft. Thanks are due to the tax advantages offered by the Internal Revenue Service. Businesses can depreciate their company planes as they can any other capital asset, earn an investment tax credit, deduct interest and taxes paid, and deduct the cost of fuel and maintenance as expenses. If they lease their airplanes, they can deduct the entire lease cost plus operating costs. Uncle Sam has made it attractive for the medium-size to large business to operate its own aircraft.

Personal flying accounts for about a quarter of the miles flown each year. Those miles are covered in every type of flying machine imaginable—from Waco biplanes to Pitts aerobatic planes to Luscombe taildraggers and Cessna 172s. With engines that are often good for 200,000 miles or more, many of the planes made since 1950 are still in the air regularly.

The reasons for personal flying are just as varied as the aircraft. Many private pilots use their wings to expand their horizons. They take their families on cross-country trips to visit friends, relatives, or Disneyland. They fly the family on vacations to ski resorts, camping spots, or beach retreats. They use their planes as they would their station wagons.

Other private pilots use their airborne toys to be alone in the sky. They take a few hours off each weekend to climb into an old T-craft or Piper Cub and inspect clouds. They shoot touch and go's at rural fields that rarely see a plane except on weekends. Their wings take them to quiet places in the sky.

Some pilots commute to their jobs in small planes. They may live many miles from their work and use that as an excuse to be airborne every morning and afternoon. By having a car on the other end, they can often get to work in half the time their co-workers need.

Still other fliers are sports pilots. They see their planes as competitive tools. They compete against others and themselves. They enter races. They fly aerobatic competitions. They leave their regular jobs and troubles on the ground while they polish their skills and strive toward personal goals.

Actually, all those are excuses for flying. The individual pilot's real reason can go much deeper. Richard Bach, author of *Jonathan Livingston Seagull,* says that he gains *perception* from flying. While he is cruising 2000 feet above the ground in a biplane that has seen most of aviation's history, he can detach himself from his troubles on the ground below. From the sky above, the problems that are so great within the walls of his home seem insignificant. They are put into the perspective of an overall view of hundreds of homes and thousands of people.

Another reason for flying is uncovered in the flight office on a cloudy day: self-discovery. Warmed by laced coffee, pilots with all kinds of experience admit that they fly because they want to prove something to themselves. They want to tackle a challenge that the majority of people would never accept: faith in their planes and in their own skills to overcome the inherent fear of flying. They want to be looked up to as brave and skillful. They climb into enclosed cockpits and surround themselves with instruments that assure them of safety, but in their minds they step into open cockpits, strap on leather flying caps, tie their long scarves around their necks, and yell "Contact!"

Flying is a vision.

The Price of Wings

Okay, back to earth. What's all this flying going to cost me?

The cliché that "if you have to ask the price, you can't afford it" can be applied to learning and enjoying flight, but it doesn't have to be. Other pilots will quote cost figures like "$38 per hour" and "$54 per hour." On talking with them further, you'll find that their costs are high because:

1. They fly complex planes.
2. They fly only a few hours a month.
3. They do both.

Flying is much like driving a car. It's costly and impractical to buy a 1980 Detroit Spitfire car with chrome-plated everything just to drive to the super-market once a week. A taxi ride is much cheaper. So it is with planes. If you use your wings for business and log many hours of efficient flying each month, you'll be able to justify a better airplane and a lower per-hour cost. But if you're spending just a couple of hours each month in the cockpit to keep your flying skills fresh, you can rent your plane more cheaply than you can own it.

In other words, an estimate of the cost of your newfound hobby involves a number of factors: How much will you use your plane? Business or pleasure? What will it cost you to purchase a plane? Maintain it? Operate it? Will you use your plane to fly long distances you now must drive? Will your plane replace your recreation vehicle? Will it provide recreation for your whole family?

Your Rich Uncle

Flying is like just about everything else—the government has its finger in it. In this case it's the Federal Aviation Administration (FAA). The FAA licenses pilots and planes and regulates the air space in which they travel. To fly an airplane in this country, you and your wings will have to be checked out by the aviation branch of Uncle Sam's family. You may not be happy about FAA's close scrutiny of your own skills as a flier, but you'll be glad that the skills of other pilots get it.

The FAA says that to fly an airplane you must have:

A Student Pilot Certificate
A Class III Medical Certificate
An FCC Restricted Radiotelephone Permit
A total of 40 hours of flying time including 20 hours of dual instruction and 20 hours of solo time

You must also be at least 17 years old and be able to read, speak, and understand English. Those are the requirements for a private pilot license. The license entitles you to fly a single-engine airplane that has been certified as airworthy by the FAA. You can fly under what are called visual flight rules (VFR), which allow you to fly when visibility meets certain minimums. Later on you can upgrade your rating to fly by instruments and fly in any weather that's safe. You'll find more about advanced licenses in Chapter 10.

Now let's talk about what a private pilot license is going to cost you. Even though the FAA usually demands a minimum of 40 hours of instruction and solo time, most students spend 50 hours or more on gaining the proficiency they need to pass the licensing test. The actual time depends on many factors that can be varied by the conscientious and cost-conscious student. We'll cover them in depth in the next chapter. For now, let's look at the average cost of instruction.

To pass the written part of your FAA test, you'll need to study rules and regulations governing pilots. You can buy the needed study books direct from the FAA for about $10; you can buy privately printed study aids that help you understand how the test is constructed and passed for $10 to $20; you can buy tapes of ground school instruction for $50 to $75; or you can attend a weekend seminar ground school for $125 to $150. Your choice of method of passing the written test depends on both your funds and your ability to study on your own. If you're fresh from four years of college, a couple of books and a few days of cramming may get you a passing grade. However, if you left school 30 years ago on a dead run, you may want to invest in the more expensive tutored ground schools. They may be the cheapest way out.

Flight instruction is obviously more costly than passing the written test.

You must rent the instruction plane, with its avgas, oil, maintenance, insurance, storage, and related costs, and also hire a competent instructor who can teach you what you need to know in the shortest time.

Rental costs of trainer planes vary greatly. They vary because of the initial cost, complexity, age, and condition of the plane, local demand, and other factors. Planes of the same model can rent for $12 or $24 per hour at airports just a few miles apart. Instructors, too, vary in cost—from $10 to $18 per hour. As you can see, shopping around for instruction can make a big difference in cost.

To answer the question how much a private pilot license will cost, let's estimate 25 hours of dual (student and instructor) flight time and 25 hours of solo flight. Let's use $12 per hour for the instructor and $16 per hour for the trainer plane; those are realistic figures for the comparison shopper. The total cost (not including ground school) comes to $1000. If you're really on a budget, you can trim that price tag even more with techniques offered in Chapter 2. As an example, though, let's say that the average student who is willing to do a little shopping around can earn a private pilot license for about $1000.

For that thousand you learn how to take yourself into the clouds—and come home again.

The cockpit of a modern private plane seems overwhelmingly complex to the new pilot because it is. This control panel belongs to the $350,000 Beechcraft Duke. Basic flying machines may have only half a dozen instruments. *(Beech Aircraft Corporation)*

Keeping It Up

Once you've earned your wings, you'll be anxious to head off "into the wild blue yonder." What's it going to cost? There are actually three costs you must consider:

Initial cost
Fixed costs
Variable costs

Each of those costs can vary greatly depending on how you obtain your plane. Here are the most popular methods:

Purchase
Rental
Lease
Partnership
Flying club
Building your own

If you purchase your wings outright, your initial cost or down payment will be about 20 to 25 percent of the price. That is, if you decide on a used Piper Tri Pacer at $9500, your down payment will be $1900 to $2375.

If you rent your wings, your initial cost is zero. Of course, your hourly rental fee is higher than the hourly cost of operating your own plane would be. Also, your favorite plane may not always be available when you're ready to fly. We'll have more on the pros and cons of renting later.

Leasing often involves a small deposit as an initial cost, but, for the businessperson, it offers the best features of purchasing and renting.

A happy medium for many pilots is the partnership. Two or three pilots go in together to buy and fly a plane. A partnership does have disadvantages; but if you select your aviation associates carefully, you can cut your flying costs dramatically while having the use of a better plane than you could afford by yourself.

A flying club is simply a partnership with bylaws and more partners. For an initiation fee of $50 to $200, monthly dues, and a reduced hourly rental fee you can select among many craft. A flying club multiplies the good and bad aspects of a partnership.

Building your wings can be both economical and fun. With an investment of $5000 to $10,000 and 1000 to 2000 hours you can build your plane —and fly it too. This is a growing option for the budget flier.

Those are the initial costs of obtaining your wings by the most popular methods. To illustrate fixed and variable costs, let's assume that you are purchasing a plane either by yourself or with others or that you are renting or leasing it "dry," that is, without fuel and insurance.

The fixed costs of flying include many things. They begin with the monthly payment or monthly dues and include all recurring costs of ownership such as licensing, insurance, storage, and required annual inspections. That is where the consumer pilot can really save, for the fixed costs remain about the same even when a plane is used more. In other words, increased utilization actually reduces fixed costs *per hour*.

Variable costs of flying go up as you do. They include:

Fuel   80/87 octane aviation fuel used in older planes costs slightly more than gas for your car. 100/110 aviation gas is even higher.

Oil   Plane engines are of the internal-combustion type, and they require oil much as your car does.

Maintenance reserve   Most aircraft engines require overhaul at intervals of 1200 to 2000 flying hours, depending on age and use. A smart aviation consumer will set aside an estimated amount for each hour flown to cover the cost of the inevitable $3000 to $6000 engine overhaul.

Landing fees   Some airports will charge you landing fees for use of their runways if you land at them. Normally this is a small cost unless you do a lot of traveling.

Spare parts   Since there are usually no aircraft parts houses on emergency landing strips, it's best to have a few of the most common spare parts on board when you fly. As they are used, they become one of the variable costs of flying.

The relation between variable costs and hours flown is exactly the opposite of that between fixed costs and hours. That is, the more hours you fly, the greater the variable costs.

A smart aviation consumer can reduce the initial, fixed, and variable costs to a total of less than $15 per hour and capture the exclusive thrill of flying on a budget.

Winging It

Well? What do you think? Does flying seem both exciting and affordable to you?

The initial cost of flight instruction is not much greater than the cost of learning to be a good skier. The hourly cost of flying can be trimmed to the point at which flying is competitive with other recreational activities while still being safe. And if you can combine business with pleasure, you can take advantage of current tax laws to reduce the costs of going up even further.

So it all comes down to the same thing: you can learn how to fly and enjoy flying on a budget if you want to. You can use smart consumer techniques to get your money's worth from every dollar you spend. You can earn your wings through the most practical and economical methods. You can discover new ways of using your plane to get both utility and value from your aviation buck. You can join the growing group of people who refuse to be grounded by the seemingly high cost of general aviation. You can be a budget flier.

Making Your Decision

Flying is paid for with both time and money. To learn to fly, you'll have to invest a total of 50 to 100 hours in actual flight instruction and practice, review, and studying for the written exam. Then once you've earned your wings, you'll want to keep your investment fresh with at least a couple of hours each month in the air.

How much time can you invest in your new hobby? Can you spend three or four hours a week for the next four to six months? Will you have time afterward to enjoy flying on a regular basis? Can you break away from your job during the week for flight instruction or practice? Budget flying means budgeting both your time and your money.

Talking about money again, you've seen what it will take to get your license—about $1000—and you've seen how much cash it will take to keep your skills current once you have that license—$10 to $25 per hour. At an average of an hour of flying per week—50 hours per year—your new hobby can be enjoyed for $40 to $100 per month depending on the plane you rent or buy and the hours you fly.

Is that within your price range? Are you willing to spend that much to learn to fly? Will you be able to set aside the cash to enjoy your hobby on a regular basis? Can you justify your investment by earning a business expense write-off, earning a few dollars with your wings, or combining flying with other recreational activities?

Flying, like any other worthwhile activity, requires a commitment. You must tell yourself you are going to do it, and you must then do it. The first part of the commitment is setting a goal for yourself.

What is a goal? It's a destination, it's a place you want to be sometime in the future. It can be an actual place, or it can be a stage in life or a level of knowledge that you want to attain.

The concept and design of the Bellanca Citabria have been enjoyed by budget fliers for over 35 years. The Citabria cruises at 115 miles per hour while getting over 25 miles per gallon. It's a well-proven plane for the flier on a budget. *(Bellanca Aircraft Corporation)*

To set a goal for yourself, you must make it:

Realistic
Attainable
Measured

That is, your goal must be one that can be reached within a specified amount of time. If you were to set a goal for yourself of being the first human on Venus within the next six months, you would never reach your goal. It isn't realistic. But if you decide to earn your private pilot license and begin flying in the next six months, you have a realistic, attainable, and measured goal, a goal you can reach.

Next comes the planning. To reach your goal within the timetable you've set for yourself, you must plan the steps that will take you to it. If you've set as your goal getting your license and starting flying within the next six months, you can develop a plan something like this:

Take an introductory flight this week to make sure that I want to learn to fly.

Begin shopping around next week for the best low-priced instruction in the area.

Check my funds to make sure that I have the money to start toward and reach my goal within a reasonable time.

Check my available time to be sure that I can schedule lessons frequently enough to retain what I learn.

Decide which flying school I will use to earn my license and begin lessons.

Once I have my license, review my budget and expected hours of usage to decide whether to buy, rent, or lease a plane, join a partnership or flying club, or build my own wings.

Discover new ways to cut my flying costs and increase the amount of sky time I can get for my dollar.

Enjoy budget flying.

Finally, it's time to take action. Don't just hope, dream, and plan your flying—get started. Take the first step in your plan and complete it within a reasonable time. Then take step two and complete it. Move up your plan until you've reached your ultimate goal to earn your private pilot license and enjoy flying economically.

Each day, thousands of average people with above-average dreams discover that there is room in the sky for the not-so-rich. They have blended the sensation of flight into their lives without taking second jobs or cashing in insurance policies. They have done it by becoming smart aviation consumers. They have discovered the friendly skies of budget flying.

2 LEARNING TO FLY ON A BUDGET

Each year about 200,000 new students take their first flying lessons. Of them, about 85 percent build up the hours and skills necessary to join the flying community as private pilots.

Many of the students—business executives, auto mechanics, secretaries, real estate salespeople, and housewives—attend manufacturers' flight schools for set fees that guarantee them their wings. Others hire their instructors and planes by the hour.

In each case there are techniques that the student can use to cut both the cost of flight instruction and the time it will take to earn a private pilot license. After all, budget flying is more than just saving money; it's also saving time. Your time is valuable too.

Learn and use the aviation consumer techniques, and you can cut the cost of flight instruction by as much as half.

The High Cost of Learning to Fly

On an average, students spend about $1600 on the way to earning their wings. That includes the cost of renting the plane, the instructor's time, ground support, and incidental fees.

Why so much? Many reasons. First, the plane that the student rents, though not fancy, is probably a newer plane. Most flying schools are also aviation dealers who buy and sell aircraft. Knowing that most students will purchase or lease a plane of the brand they learned on, flight schools–plane dealers make sure that student planes are the newest and easiest-to-operate planes available. Since most trainer planes today cost from $20,000 to $30,000, the hourly cost of trainer rental is higher than it would be for an older plane. You're renting their floor models.

Maintenance is also expensive. New or not, all planes require a certain amount of maintenance because of the high number of hours they are flown and the rougher treatment they get as student planes. Nearly all maintenance must be by FAA-licensed airframe and power plant (A&P) mechanics at $25 or more per hour. Parts also are more expensive in an industry that turns out 15,000 new planes a year compared with Detroit's 9 million units.

The cost of aviation fuel has risen as fast as, if not faster than, that of automotive gasoline. The pump price has more than tripled in six years, and there are threats of future increases. Even though most trainer planes get better gas mileage than the family car, the increasing gasoline cost is a major factor in the increase of instruction rates.

Renting the instructor's time takes up a third to a half of your learning-to-fly budget. Instruction fees range from $10 to $18 per hour for the private pilot license, and they are even greater for higher-class licenses.

Why are they so costly? First, the instructor doesn't get all that money. As much as half of it goes to the school where he or she works. And the wages must cover both the time in actual in-flight instruction and the time between students. The typical instructor earns less than the students.

Your instructor has compensations other than money: earning a living by doing something enjoyable while building time toward a higher rating. While you are logging your time, your instructor is logging time also. That contributes toward eligibility for licenses necessary for a job as an airline transport or other commercial pilot.

Ground support offered by the flying school also takes up some of your flying money. The cost of instruction includes office overhead, wages for the chief instructor or administrator, records and paperwork needed, cost of rent, lights, and advertising, franchise fees, and training aids. It all adds up.

Another, but indirect, cost of learning to fly is one that you yourself can control: your ability to learn. Since only a few students ever complete instruction within the mandatory 40 hours, the cost of instruction is directly related to how well you learn and how much you retain. The faster you learn, the fewer hours you have to purchase.

Later in this chapter I'll give you 10 proven ways to make you a better student pilot and cut your instruction costs. For now let's look at the three primary reasons why people learn or don't learn.

The first is *concentration*. Because of the increasing complexity of life, it's often difficult to concentrate on any one thing for long. While you're working or driving, you may find your mind wandering to a dozen other elements in your life. Of course, the first day at your job or behind the wheel of a car you used all the concentration you could muster. You were learning something new, and you reminded yourself you had to retain it. Learning to fly is like that. You are acquiring new concepts and new ideas. You must concentrate on what the instructor is saying and what you are doing. Save the sightseeing for after you solo or earn your license.

The second element of learning to fly in the shortest time is *retention.* That is, you must not only hear what is being said, you must also remember it. Your instructor will help you with this by questions and summation as you return to the airport and on the ground after the lesson. Later in this chapter I'll give you a couple of easy ways to remember what you've learned and cut down the review time you need during each lesson.

The third element of learning is *motivation,* or attitude. If you're a willing student who is sincerely interested in flying and learning as much as possible in the shortest time, you will learn much faster than the student who isn't excited about flying. A limited flying budget will also help motivate you to learn how to fly in the shortest time.

Shopping for Instruction

You have your choice of over 4000 flight schools. Which one should you choose to show you how to fly on a budget? The answer is as individual as the student. Flying schools come in all shapes and sizes—from the one-plane one-pilot grassfield school to the modern multimillion dollar campus of an aerodynamics university. Bigger is not necessarily better. The instruction you get from the grease-knuckled owner of a grassfield school may be as good as or better than that from a big-city flight school.

Buying instruction is much like buying any other big-ticket item: it takes some shopping around. Here's where to look:

Dealer schools Most general-aviation manufacturers offer flight training through their dealers. In the Yellow Pages of any metropolitan area you will find ads for Beech Aero Clubs, Cessna Pilot Centers, Piper Flite Centers, and others. In most cases the training programs are developed by the same company, Jeppesen Sanderson, and the difference is in the plane and the packaging.

Independent schools Many fields offer flight instruction that is independent of any manufacturer or dealership. Such schools are not selling particular models of new planes. Their equipment is often older, but it is in good repair and can usually be rented at lower rates than that at dealer schools.

Colleges and universities If you're working toward a career in aviation, there are a number of good resident schools where you can learn to fly for pay. The biggest are Embry-Riddle Aeronautical University, Spartan School of Aeronautics, and Ross School of Aviation. A current copy of one of the flying magazines will give you the addresses of these and other pilot schools.

One of the differences you'll find in flying schools is that some are "FAA approved" and some are not. That is no reflection on ability to teach students to fly safely. The main difference is that the curriculum at a certified flight school has been approved by the Federal Aviation Administration. The real advantage to the student is the number of hours of instruction required. A student at an approved school must have a minimum of 35 hours of instruction, 5 of which can be in a lower-cost flight simulator. Instruction in nonapproved schools must be at least 40 hours.

That doesn't change things much for the average student, who will spend over 50 hours of flying time to get a license, but it does offer a small savings to the student who already knows flight concepts and can learn quickly. Of course, most FAA-approved schools must charge more per hour than nonapproved schools to cover their cost of certification, but there is potential savings for some students if they use an approved school.

The biggest difference between schools lies in the individual instructors and the value of the planes the school uses as trainers. To teach flying and sign a student's logbook, the instructor must be a certified flight instructor (CFI) and have hundreds of hours of flying time and be able to pass exacting written and in-flight tests.

When shopping for a specific flight instructor—one in whom you're willing to invest a thousand dollars and with whom to entrust your life—you should look for two important things: experience and attitude. A good CFI may have attended one of the aeronautical schools or may have learned at a rural grassfield school. What came next? Some special flying courses? Bush pilot school? Aerobatics school? Mechanics school? How many hours of flying has the CFI? How many months or years as an instructor? As for attitude, why does the CFI fly? Is instructing a planned and agreeable future? Or is it just a way station until the hours necessary to qualify for an airline job are logged?

While you're checking out flying schools, ask about the trainers that are used. Since most trainer planes have very similar instruments and handling characteristics, your biggest concern should be for safety. How are the planes maintained? Is insurance included in the rental fee? How many hours do they have on them? Age, you may be surprised to hear, has little to do with safety in planes. One successful flying school uses six Cessna 150s ranging in age from 5 to 15 years and rents them for just over half of what the new Cessna 152s rent for. They are kept in good repair and must be reserved a week in advance because of the demand.

Another way to test a flying school or independent instructor is to ask for the names of a few students who have passed the private pilot test successfully. Talk with those students. Did they feel comfortable with the school and instructor? Why or why not? How many hours did instruction take? What was the total cost? Are they satisfied with the results?

Finally, ask the school about its ground school, if any. Some dealership schools offer modern audiovisual training aids to augment and reinforce the student's acquired knowledge; others rent their instructors on a one-to-one basis to help the student review FAA regulations and study aids. No one system is best. All are designed to do the same thing: give the student the greatest knowledge of flying and regulations in the shortest time.

"How Much Is This Gunna Cost Me?"

Once you've settled on two or three of the best schools or instructors, it's time to talk price. To make the best decision, you must weigh both the cost and the value. Will you be paying for instruction by the hour, or will you pay a package price that guarantees you will get your license? If it's a package, how many hours is it expected to take you? Is ground school included in the package? If not, how much extra will it cost? If instruction is by the hour, how many hours does the school estimate it will take you to solo? To earn your license? How busy is the school? Can you schedule lessons when they are convenient for you, or will you have to take time off from work?

Shopping around can make a big difference in the total instruction price. As an example, in a recent survey taken near a metropolitan area it was found that rental fees for the same trainer plane—a Cessna 150—ranged from $12 to $22 per hour for solo and from $22 to $34 for dual (with instructor) operation. The package plans in the area ranged from a high of about $1800 to a low of $925 plus a few dollars for books.

The points to be stressed are that neither the highest nor the lowest price is necessarily the best value and that the best value will be found by comparison shopping. You may decide that the best instruction you can get for your dollar is from an FAA-approved dealership school, or you may find a CFI who owns a plane and will use it to teach you to fly on his or her days off from a regular instructor's job. In any case, to make the best decision of value for your flying dollar, you need to understand why instruction costs are comparatively high and what you can do about it.

Ten Ways You Can Save on Instruction

How much you get out of your flight instruction depends on just one factor: You. You are the variable that can make the biggest difference in the cost of your flying instruction. Why? Because the faster you learn the less you'll have to pay for lessons. To help you learn the most for the least, here are 10 proven ways to keep instruction costs down.

Be prepared. The Boy Scouts have the right idea. By being prepared for your lessons you'll be able to retain more of the new concepts and techniques that your instructor will be firing at you.

The first segment of preparedness is knowledge. In each of your lessons you'll be reviewing things you've already learned and then learning new concepts about flying and putting them into practice. To be prepared for your lesson, ask your instructor what you'll be covering in the next lesson and then spend time studying it. In the first few lessons before you solo and get your student license you'll learn about preflight actions, straight-and-level flying, climbs, descents, turns, slow flight, stalls, ground reference maneuvers, and landings. If you've read up on those techniques just before your lessons, you'll catch on to what the instructor is teaching you quicker and save costly review time.

Preparedness also means attitude. What you're learning during your lessons may save your life and should be taken seriously, but most students are so intent during the first lessons that they miss the fun of flying. Part of being prepared for your lessons is reminding yourself that safe flying can be fun.

Be receptive. Flying will be completely new to you. You've driven cars and possibly other vehicles, but the concepts and skills of flight are something else. To understand them and to make them part of your subconscious, you must be receptive to new ideas and techniques.

One highly successful instructor helps his students become more receptive and open to new ideas by helping them clear their minds when lessons start. As they first sit in the plane and tighten their seat belts, he reminds them that they can leave their problems on the ground and pick them up when they land. Years afterward, his former students remember to "clear their minds" of everything except flying when they fasten their seat belts.

Once your mind is clear of the outside world, you must concentrate on the job at hand. You must remind yourself that you are flying and that everything must be done slowly, deliberately, and thoughtfully.

Relax. You will be ·nervous during instruction; every student is. You *should* be nervous. Flying isn't one of the human being's natural skills. The idea, though, is to control your nervousness and channel that energy into awareness and concentration. You do that by keeping your mind and body in a state of relaxation as much as possible. You start with making sure that you are comfortable in the pilot's seat, that your belt is adjusted snugly, and that your feet are comfortable on the rudder pedals on the floor.

Relaxation must also carry over to your mind. When you find yourself becoming tense or nervous as you fly, stop and relax your shoulder and back muscles and loosen your grip on the control wheel. Clear your mind of doubt and worry. Remind yourself that your instructor is right beside you and is guiding you. Trainer planes are really very docile; it is difficult to get them into a spin or other dangerous maneuver. Even if you stall it, the plane will recover itself. So relax and enjoy the flight.

Be retentive. There are only a handful of maneuvers that you will learn in your training for a private pilot license. Each maneuver will be repeated and practiced over and over until you know it automatically.

You can cut review time dramatically by learning the requirements for each of the maneuvers as quickly as possible and retaining them in your mind. Your instructor will help you with acronyms. CIGARS, for example, will remind you that, before taking off, you must check controls, instuments, gas, airplane trim, run-up, and safety. There are acronyms for wind correction angles, landing duties, and other tasks. Write them down. Learn them. Review them. Practice them at home. Put them into your mind to be drawn out as they are needed.

Sometimes things will seem to be coming at you too fast. As soon as you are introduced to a concept and think you understand it, a new element will be introduced and you may be confused. Then learning will halt. The cure for confusion is to stop what you're doing and thinking and regroup. Have your instructor go over the concept and application again. Make sure you see how it works and how it fits into the job of flying an airplane. Then move forward. Build your knowledge and skills on a solid foundation of tested and related facts of flying. You'll save a great deal of time later.

Ask questions. The instructor you hire to teach you to fly is your employee, one you've hired to pass on knowledge and skills and help you develop flying skills of your own. Sometimes in this relationship an important idea is lost. You may be saying "uh-huh" while actually concentrating on the horizon indicator or wing tip. The instructor may be covering flight procedure before explaining and illustrating the concepts. In any case, an important idea may be lost to you if it isn't questioned.

Whenever a question arises in your lesson, ask yourself whether it is relevant to what you are doing right now. If it is, ask it. The answer may help you put the concepts of flight deeply into your mind. It may clarify a misun-

Today's most popular trainer plane is the Cessna 152. *(Cessna Aircraft Company)*

derstanding you have about a maneuver. It may save hours of instruction time.

Other questions arise between lessons, and have to be answered. The smart student carries a small pocket notebook in which to write down questions for the next flying lesson. Some questions can be answered by reviewing the flight books, but others can be answered only by the instructor. You should ask for a few minutes before and after each lesson to feed questions to your instructor. That will help you understand flying and help your instructor discover how much you are retaining and what areas need more work.

Keep a journal. Here's one idea that has helped many students retain knowledge between lessons—especially when the lessons are a week or more apart. It starts with a simple notebook in which the student can write down what was done and learned during each lesson as soon after the lesson is completed as possible. A sample portion of a student's journal might read as follows:

> Lesson 2 in Cessna 44-V with instructor Mark Johnson for 1.3 hours. Did preflight from memory and went through run-up sequence with CIGARS. Hands off control wheel when taxiing. Use rudders, brakes, and throttle only. Taxied to runway 28. Aimed straight along runway and watched out side window for reference. Full throttle. Slight right rudder. Back on control wheel (elevator). Keep back pressure on wheel until plane lifts off. Maintain climb attitude to 900 feet, then coordinated left turn (ailerons and rudder). Pull out of traffic pattern at 45 degree angle.

Your journal entry should be made soon after your lesson to ensure that you include all the things you did during your lesson. You have two reasons for keeping a journal: to record and to review. As you write down your actions during the lesson, you are clarifying those actions so that you understand them better. Later, you can review your lessons and refresh your memory before the next lesson. More on this later.

Do your homework. With your practice sessions you must have flight theory. You must know and understand the four forces that are at work when a plane flies. You must know the rules and regulations that govern pilots. You must discover the purposes behind the maneuvers you learn.

That is what ground school is all about. Whether you decide to study on your own, with an instructor, or in a weekend seminar, you must learn the elements that will be on your FAA written test for private pilots. The best way to study alone is to set aside at least two hours each week—and preferably more—when you can sit down and go over your lessons and possible test questions without interruptions. But a study period should not be more than two hours long. Each period should include a review of what you did during your last lesson as well as a study of maneuvers you will practice in your next lesson.

Use your imagination. One of the best study aids used by the budget flier is imagination. With it the flier can take off and land, practice S turns and wind correction angles, and set plane attitudes firmly in mind—free.

During your study period you can sit back in a quiet place and review all the maneuvers you practiced during your last lesson. You can see yourself making a smoothly coordinated climb—oops, more right rudder—or practice stalls without disturbing your stomach. Instructors know that this type of review is often as valuable as actual flying time because it allows the student to stop the action and put it on instant replay.

Some students help their imaginations by getting photos or posters of the control panels in their trainer plane. (The photos or posters are available from the manufacturer or through publishers who advertise in flying magazines). With a photo of a panel tacked onto a wall, a student can take time to become acquainted with instruments and their locations, as well as to review flight maneuvers without having to rent a plane and hire an instructor.

Hang around. Here's another enjoyable way to add to your knowledge of flying at no cost: become an airport bum in your spare time. Once your lesson is over or even when you have a spare hour, acquaint yourself with the sights and people who make up a general-aviation airport. Spend time in the flight office reading flying magazines and literature and talking with other students and pilots. Walk the field while inspecting planes and acquainting yourself with their features from outside the windows. Watch pilots take off and land. Watch how they fuel up and preflight their own planes.

Another benefit of airport bumming is that you may find a pilot who is going up for an hour or two to keep skills sharpened and would enjoy your company. You will be able to see another pilot at work. Don't let another pilot's methods undermine what your instructor is teaching you—your instructor probably knows best—but do try to pick up the pilot's skills by watching closely. You may find a technique or concept that will help you understand flying better.

Spare time at the flight office and airfield can give you knowledge of and insight to the new world of flying.

Schedule lessons. The best is last. The tenth way to keep instruction costs down is to schedule your flying lessons so that you get the most good out of them. Even with journals and prelesson reviews, your retention of flying concepts and techniques can fade if lessons are a week or more apart. Experienced instructors say that the best schedule is two or three lessons of an hour and a half each per week. That schedule means that there are only two or three days between lessons, so the student will retain much of what he or she learned during the last lesson. Less time will have to be spent in review.

The hour and a half lessons are best because they allow time to cover new flight knowledge before heading back to the barn. The normal one-hour lesson gives time for a preflight inspection, a flight to the practice area, a short review, a short introduction to new material, and then time to return. The 1$^{1}/_{2}$-hour lesson usually allows 45 minutes or more to cover new material. Some instructors ask for 2 hours: 15 minutes on the ground prior to takeoff, 1$^{1}/_{2}$ hours in the air, and another 15-minute period for review and questions on the ground.

Of course, many time clock workers might think that three lessons a week are out of the question. They're not. Many students plan one lesson on Saturday morning, one on Saturday afternoon, and one Sunday afternoon (or other days off). Others plan two lessons on the weekend and one right after work midweek depending on the time of year and sunset.

Another method of scheduling lessons effectively is to plan to earn your license during an extended vacation period. One experienced instructor, now an FAA safety officer, helped a student earn his private pilot license in just 21 days by scheduling morning and afternoon lessons. Not only was that shorter than the standard four to six months at two lessons per week but it also took the student fewer hours to earn his license. He passed his test after just 35 hours at the FAA-approved school. Although the schedule and results are exceptions rather than rules, the instructor felt that the twice-a-day schedule dramatically cut the number of hours the student needed by keeping concepts fresh in his mind and cutting review time.

Cockpit of the Cessna 152. Pilot's seat is on the left (in front of primary instruments); navigation and communication radios are in the center; and copilot's seat and secondary instruments are on the right. You can cut the cost of flight instruction by practicing with your imagination and a photo of your trainer's plane. *(Cessna Aircraft Company)*

Flying Lessons for the Frugal

The cost of flight instruction is gaining altitude. Planes are expensive. Aviation fuel and maintenance costs are high. Instructors are costly. Even so, there are many ways in which you, the aspiring pilot, can cut the cost of lessons and earn your private pilot license safely and inexpensively:

Be prepared for your lessons.
Be receptive to new concepts and skills.
Relax and enjoy the flight.
Be retentive and memorize facts you need.
Ask questions of your instructor.
Keep a journal of your lessons.
Do your homework and study flying rules.
Use your imagination to take you flying free.
Hang around the airport and learn.
Schedule frequent lessons to learn faster.

3 CHOOSING YOUR WINGS

You are not, by nature, supposed to fly.

And yet you can feel comfortably at home thousands of feet above the bird's territory while reviewing the individuality of clouds. You can share the revelations of flight with others who join you in your mechanized journey. You can ignore the elements that ground flying creatures and spend hours in the air without touching down.

But your flight is not the seemingly effortless travel that the birds enjoy. The birds can take off and land in a short distance. Fuel is nearly everywhere. Maintenance is automatic. Their navigation and communication systems are simple but adequate. You must go through a preflight inspection, estimate ground speed and calculate wind triangles, remember to retract landing gears, check the manifold pressure, turn on beacon lights, and remember to buckle your seat belt.

On the other hand, you can actually decide what type of bird you want to be. You can be fast or slow, high- or low-altitude, long- or short-distance, high- or low-wing, blue or red. You can choose your own wings to fit your needs and desires.

You discover those needs and set your flying priorities by reviewing the ways you will use your wings. Business or pleasure or both? Economy or time-efficiency? New or used? Short takeoffs and landings? Unimproved field landings? Water landings?

Other considerations in choosing your wings include seating, available load, power, fuel economy, cruise speed, range, wing configuration, landing gear type, propeller type, avionics, and other special requirements.

Finally, as an aviation consumer, you must review your own flying budget, the methods of ownership, maintenance considerations, operating costs, and resale value.

Why so many decisions? Because our technological and economic system offers a wide variety of choices to fit our individual needs. Some pilots make speed and comfort their top flying priorities; others place economy at the top of the list. No one plane can please everyone. In fact, no one plane will be exactly what any pilot is looking for. It will be a compromise. The best compromises between flier and wings are made by aviators who know both what they want and what is available.

The Informed Compromise

Many pilots are looking for the simplest planes they can find to take them to the clouds and back. They're not going for any speed or endurance records. They fly to get away rather than go.

That type of pilot is looking for what's called a *low-performance plane:* a single-engine plane rated at under 200 horsepower with fixed-pitch propeller and fixed landing gear. Very simple.

The low-performance plane is built for power efficiency. It's lightweight and uses minimum fuel; often it gets more than 20 miles per gallon. It's built and operated simply. That ensures a lower initial price tag and lower maintenance and fuel costs.

The low-performance plane is often chosen by the pleasure pilot or short-distance business pilot, who may fly only a few hours a month in a local traffic pattern or up to a thousand miles away. Longer trips in a low-performance plane are often uncomfortable because of space and comfort limitations. The majority of budget fliers choose low-performance airplanes as their wings because of the efficient use of fuel and the lower initial cost.

However, some pilots see economy and utility in high-performance planes. A *high-performance plane* is a single-engine craft that is rated at over 200 horsepower. It has a controllable-pitch propeller and retractable landing gear. A controllable-pitch prop offers better efficiency when the plane is climbing or descending because most fixed-pitch props are designed for greatest efficiency when the plane is cruising. The retractable landing gear allows the pilot to pull the wheels up when in flight and thus cut down the plane's wind resistance and increase efficiency. High-performance planes are more time-efficient than money-efficient. They get you there faster.

The problem is that high-performance planes cost more than low-performance planes; they cost more both to buy and to operate. Whether they are efficient enough to qualify for your budget depends on many factors: how much flying you plan to do, whether time or cost is most important, whether you plan to use your plane for business, how many passengers you expect to carry, and what you can pay initially and per hour.

One of the most important basic instruments in flying is the tachometer. It displays the engine's revolutions per minute for efficient operation. *(Wag-Aero, Incorporated)*

Of course, if you do set your sights on a high-performance plane, there are many things you can do to compensate for the higher costs. You can purchase the plane in partnership with others, join a flying club offering high-performance planes, or rent a plane as you need it. This is another pro-con decision that depends on your intended use and your flying budget. More on buying, renting, leasing, and flying clubs in coming chapters.

Odd Birds

You may have watched birds take off in short distances, from grass and rough terrain, from snow, and even from water. Airplane designers have always wanted to copy the birds.

Their first success came in developing planes that could perform short takeoffs and landings (STOL). They simply modified their planes for STOL by streamlining the ailerons and wing tips, changing the engine horsepower, and replacing the fixed-pitch propellers with constant-speed props. Specific modifications depended on the plane model.

For their efforts, the modifiers got planes that could take off and land on shorter fields, climb quicker, and respond faster than factory models could. One STOL modifier, Robertson Aircraft, claims that insurance premiums from a major underwriter are reduced by as much as 35 percent for STOL-equipped planes because of increased safety. The underwriter found that there were fewer accidents of lesser severity with STOL than with conventional aircraft. Another plus.

Also available for the active flier are skis and pontoons that will permit landing on snow or water. Attaching floats to your plane requires a *Supplemental Type Certificate* (STC); and if you carry passengers or fly for hire, you'll need an *airplane single-engine sea rating* (ASES). That requires about five to ten hours of instruction beyond your private pilot license. In areas with an abundance of lakes, a seaplane rating can open up thousands of miles of new runways to you.

Getting More Specific

Now that you've chosen the *type* of plane that best suits your needs and flying budget, you must narrow the field down further to decide on the best wings for you. The points you should consider are:

Seating
Load
Power
Fuel economy
Cruise speed
Range
Configuration
Landing gear
Propeller
Avionics

You must also consider your own flying budget in comparison with the initial costs, maintenance costs, operating costs, and resale value. Each of those factors is important not only in getting your money's worth but also in getting a plane that is best suited to your flying needs and desires.

Stuffing Your Bird

Planes are sensitive to the amount of weight you ask them to carry. You can pack your station wagon until sleeping bags are sticking out its windows and it will still get you to your destination. But your plane has a load limit. It can carry so many people and/or so many pounds of baggage. Any more and you're asking for trouble.

As long as you remember that fact of flying you'll have no problem. The owner's manual for the plane you're buying, renting, or borrowing will offer a *maximum useful load* figure, which is the difference between the weight of the plane empty and how much the plane is built to carry. When estimating your load, add up the weights of all passengers, add the fuel at an estimated six pounds per gallon, and add the weight of any luggage or suitcases aboard.

Can the plane you're considering safely carry the weight you will ask it to carry? How many people will you have aboard? One, two, four, or more? How much will your full fuel tanks weigh? How much additional weight will you carry? Camping equipment? Tools? Equipment? Sample books? Get it all down on paper to estimate both the number of seats you'll need in your plane and the useful load you will expect the plane to haul for you. You can use both the seating and the useful load figure to help you narrow down your choices of airplane.

More Power to You

The amount of power offered by the airplane you fly also is important to you as the pilot. More horsepower simply means more get-up-and-go—and more fuel needed. Another trade-off.

Small-aircraft power plants are internal-combustion engines that work on the same principle as the engine in your car. The similarity ends there. Aircraft engines are more fuel-efficient, more dependable, and more costly than their grounded brothers. Many plane engines have an estimated time between overhauls (TBO) of 2000 hours. At an average speed of 100 miles per hour, that's an expected life of 200,000 miles.

When you are shopping for plane power, you're actually shopping for horsepower. Aircraft engine horsepower can range from 65 horsepower for a small Ercoupe to over 300 horsepower for the largest Cessna, Piper, and Trident single-engine piston planes. The horsepower rating is the maximum takeoff power delivered by the engine. Actual cruise power is 65 to 75 percent of that figure.

Another good indicator of the power offered by your plane is *power loading,* that is, the amount of weight each horsepower is asked to carry at full load. The power-loading figure is calculated by dividing the gross weight by the rated horsepower. The Piper Cherokee 140, which is powered by a 150-horsepower engine, has a gross weight of 2150 pounds. The power loading is $2150 \div 150 = 14.3$ pounds per horsepower. That figure can be compared to the Cessna 172's 15.3 or the Beechcraft Sundowner's 13.6 pounds per horsepower. The power-loading figure is not the ultimate answer

The basic navigation receiver picks up the signal of a VOR (VHF omnidirectional range) station and guides you to it. *(Collins Division of Rockwell International)*

to which plane to purchase, but it will give you one piece of the aviation puzzle.

Fuel Economy

Airplanes can be more fuel-efficient than cars. As an example, the new Taylorcraft F-19 cruises at 115 miles an hour with two aboard and uses just 3.6 gallons of gas per hour. That's 32 miles to the gallon. Nothing fancy. The Taylorcraft design and Continental 100-horsepower engine have been around for many years making auto manufacturers' claims look silly.

To estimate the fuel efficiency of any plane you are thinking of buying, simply find the *cruise speed at the 65 percent power* figure and divide it by the *fuel flow at 65 percent power.* You now have the *mpg at cruise speed* for that particular model. As the car manufacturers like to say, "your mileage may vary," but the figure is a good way to comparison-shop for fuel-efficient airplanes.

Another thing to remember is that fuel efficiency normally means engine simplicity, which is directly related to lower maintenance costs. A fuel-efficient plane will usually cost you less to feed and fix than a gas-guzzler will.

The communications radio is both a transmitter and receiver that allows you to talk with control towers, flight service stations, other pilots, and emergency services on the aviation communications band (118.0 to 135.975 megahertz). *(Collins Division of Rockwell International)*

Cruising Along

Your decision on which plane to pilot will also depend on how fast you want to get where you're going. For some pilots nothing less than the Mooney 201's 180 miles per hour will do. For others, the Bellanca Scout's 115 miles per hour is sufficient.

Most plane builders now estimate cruise speed at 65 percent of the maximum power setting while flying at 8000 feet altitude. High-speed cruise is 75 percent of full power, and economy cruise is 55 percent. In any case, it's best to use the same figures when comparing planes, and most shoppers use the 65 percent cruise figure.

When estimating the cruise speed you want in your plane, consider where you'll be going and how quickly you want to get there. If you're a weekend pilot who practices maneuvers within 20 miles of your home field, cruise is less important than rate of climb and other performance figures. If you plan to fly off to your favorite ski area 600 miles away each weekend, your time and cruise speed are more important. If you're a businessperson who must make many long-distance stops in the shortest time, your cruise speed becomes very important as you consider your wings.

So does range. If you stay close to the home field or enjoy field hopping, the number of miles to a tankful is less important than it is to the long-distance businessperson. The airplane's range will be stated either in miles or in hours in the air at cruise speed (which can easily be converted into miles). Most single-engine planes have a range of between 400 and 1000 miles with a 45-minute reserve. The range of most planes can be extended by installing wing-tip tanks. Again, it all depends on what you need from your plane.

Where to Place Your Wings

In shopping for the best plane for your flying needs and budget you've probably noticed that some planes have wings stretching from the top of the cockpit and others from the bottom. Ask the salesperson or owner why and you'll get a biased discourse on whatever type the plane happens to be.

The fact is that both high- and low-wing-configuration planes have their advantages and disadvantages. The final decision is often based on personal taste or specialized use. The high-wing plane offers better ground clearance and visibility for rough-field landings, is easier to board (especially in the rain), and doesn't require a fuel pump to get fuel from the wings to the lower engine.

Low-wing planes offer better ground stability owing to the lower center of gravity, better visibility in turns, greater ease of filling fuel tanks, and better crash protection.

The battle goes on. Nearly all Cessnas, the Maule, and Taylorcraft planes are high-wingers, and almost all planes made by Piper, Beech, and Mooney are low-wing planes. Each type is best—for specific applications and pilots. Your decision depends on your flying needs and preferences.

Landing Gears

The landing gear is a very important part of your plane. It helps you make the transition between air and land. The simpler light planes offer one of three types of landing gear systems:

Fixed tricycle gear with two wheels below the cabin and one under the nose. The arrangement provides for better pilot visibility when taxiing.

Fixed tail gear or taildragger with two wheels below the cabin and one under the tail of the plane. Practical for unimproved field landings. Lower cost.

Retractable gear system; the wheels can be drawn up into the body of the craft after takeoff. More streamlined for less wind drag and greater efficiency. Higher cost; more maintenance.

The largest number of newer light planes today are tricycle-geared in the lower price range and retractable-geared in the higher price range. Only the most economical new planes (Taylorcraft, Piper Super Cub, Bellanca Citabria, Scout, and Decathlon, and the Maule Lunar Rocket) plus many decade-old planes (Aeronca Champion, Cessna 120, 140, and 170, Luscombe, Stinson, and others) use the tail gear. The taildragger is more awkward on the ground than tricycle gear planes are, but the lower cost can sway the budget flier.

Propping Yourself Up

The propeller is another feature that can be decided by either luck or choice. It is less important as you buy or rent your plane, but it can be important as you modify a plane for efficiency. It's also a good idea to know something about the prop on the plane you're flying.

The most common type of propeller is the fixed-pitch: the angle at which the prop cuts through the air to give your plane thrust is fixed and cannot be changed. Most fixed-pitch props are cruise props that offer the greatest power, or revolutions per minute (rpm), when the plane is cruising straight and level. They are less efficient when climbing. If takeoff or climbing is more important, the plane owner can replace the fixed-pitch prop with a climb prop and lessen cruise performance; another option is to install a variable-pitch or constant-speed propeller.

The variable-pitch prop allows you to change the angle of the propeller for cruise or climb while on the ground. The more sophisticated—and more costly—constant-speed prop has a governor that automatically changes the prop pitch to maintain the engine rpm at a constant level for better performance in both climbing and cruising.

Basic Avionics

Many of the post–World War II light planes were built in the expectation that returning military personnel would turn in their cars for airplanes. That never happened, of course, but the planes of that era and the decade beyond had one thing in common: they were simple. Often two-seaters, they carried only the basic flying instruments and rarely offered a radio for communications.

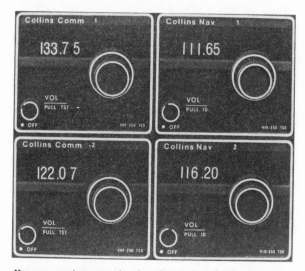

Many newer and more complex planes have both a primary and secondary
NAV/COM radio to ensure uninterrupted communications in IFR conditons.
(Collins Division of Rockwell International)

Times have certainly changed when the cost of installing avionics (that's
*avi*ation electr*onics*) in an older plane can approach the original cost of the
plane. The minimally equipped plane today usually has these three pieces of
radio gear:

NAV/COM radio VHF communications transmitter and receiver for talking with air traffic controllers and other pilots, combined with a VHF
omnidirectional range (VOR) navigation system that can help you fly right
to where you want to go.

Transponder A small transmitter that can automatically identify you on
the screen of an air traffic controller.

ELT The emergency locator transmitter (ELT) is so built that the impact of
a crash will set it off and radio your position to others.

You can operate your plane without those little black boxes, as they are
often called, but it would be like taking a trip in your car without seat belts, a
road map, and a spare tire. If the plane you are considering doesn't have
them, you can purchase new or used basic avionics units and have them
installed.

If you're a gadget collector, you can find dozens of other black boxes for

NAV/COMs come in all shapes and sizes. This basic unit can be purchased for $1100 new. *(General Aviation Electronics)*

your plane that will do everything for the pilot but replace the flight attendant. They include the following:

The autopilot can do everything from keeping the wings level to changing course and tracking a VOR radial to your destination.

Glide slope receivers give you visual indications of whether your plane is high or low in its approach to land.

The automatic direction finder (ADF) receives radio signals from transmitters and shows you which direction they're coming from. By pointing your needle in the right direction, you'll eventually arrive at the source of the signal.

Distance-measuring equipment (DME) tells you the distance between you and the signal received by the ADF, usually from a V.OR ground station.

Weather radar gives you a visual outline of the weather on a television-like screen in black and white or color.

There are more—area navigation systems (RNAV), flight directors, collision avoidance systems, horizontal situation indicators (HSI), omnibearing selector display (OBS), marker beacon receivers, and others—for the super-rich pilot who enjoys high-priced toys. For the rest of us, the budget fliers, basic instrumentation and maybe a NAV/COM, transponder, and ELT will do.

Dollars and Sense

Again, the choice of the best airplane for you must be made with your own flying budget and uses in mind. You must decide what your initial investment in flying instruction must be and then come up with an amount that you can invest in your plane if you decide to purchase one. Finally, you must know about what it's going to cost you to get yourself up into the air and keep you there.

We covered the cost of instruction in Chapter 2. The next three chapters, 4, 5, and 6, will help you make the best decision on whether to buy, rent, or build your wings. You'll learn more about the different ways of climbing into the clouds:

Full ownership
Partnerships
Flying clubs
Hourly rental
Term leasing
Building from plane kits
Rebuilding planes
Designing your own plane

We'll get into how to shop for a plane, inspect a plane for sale, and preflight a rental plane and how and where to buy airplane kits. We'll give you tips on joining the right flying club, cutting rental costs, and taking other actions that will help you as an aviation consumer.

The next chapter, 7, will cover plane operating costs and how to reduce them to a minimum safely. You'll learn:

The transponder is required if you do much flying around controlled airports. It identifies you to the control tower on the ground. *(General Aviation Electronics)*

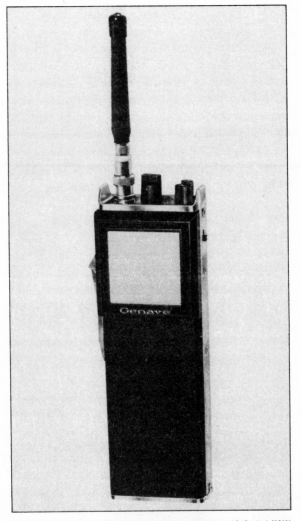

Many economy-minded fliers who own planes without sophisticated NAV/COMs have found the hand-held communications transceiver an excellent investment at under $500. *(General Aviation Electronics)*

How to buy low-cost hull and liability insurance
How to improve fuel economy in your plane
How to cut storage costs for your plane
Where to hangar your aircraft
How to reduce yearly and hourly operating costs

Chapter 8 will help you reduce the costs of plane maintenance. You'll discover which repairs you, the owner, can legally make to your plane and which you can't. You'll see how thousands of plane owners legally sidestep FAA rules and do all of their own repairs. You'll learn about the importance of FAA Airworthiness Directives (ADs) and how you can save thousands of dollars as a plane buyer or owner by watching for those repair notices. Most important, you'll learn how to cut maintenance costs while keeping your plane safe and economical.

Chapter 9 will show you how to upgrade your wings on a budget, Chapter 10 will discuss how to add advanced ratings to your private pilot certificate, and Chapter 11 will offer a number of money-saving ideas on how to make your flying pay for itself through tax savings and part-time business ventures.

Each of those chapters will help you choose the right airplane. To help you even more, two special chapters have been added. They offer a dozen each of Best New Aviation Buys and Best Used Aviation Buys. They have photos, performance facts, and indexes of utility and economy. They can help you immensely in the difficult task of choosing your wings.

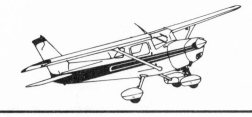

4 BUYING YOUR WINGS

Possession is nine-tenths of the fun!

At least that's what many owner-pilots will tell you. They enjoy the power of having their own aircraft always at their disposal. They fly their personal planes every day—in their imagination if not in fact. They like the feeling that comes with owning their own planes.

On alternate days those same owner-pilots curse the high price of hangaring, the creeping loss from depreciation, and the daily costs that go up even when their planes don't.

What's the answer for the budget flier? Should you buy your own plane or shouldn't you? For some pilots, plane ownership is a real bargain; for others it's a very expensive hobby. Your decision to buy or not to buy must be based on facts more than feelings if you hope to join the friendly skies of the budget flier.

Flying Facts

Before you decide whether you want to buy a plane, let's take a look at how and why to do it.

The selection process begins with a search of the many sources of for-sale planes. Surprisingly, there are probably hundreds of planes on the market in your area. As you learn the shorthand that pilots use to describe their planes' features, you'll be able to match those features with your needs and so narrow the field.

Then you can check out each plane individually to see whether you have found the best buy for you. You'll make an initial inspection and a flight inspection, review logs and Airworthiness Directives, have the plane checked out by a mechanic, and then sit down to estimate the operating costs and whether the purchase will fit your budget and your needs.

You'll pay special attention to the plane's initial cost and how you plan to finance the purchase. You can save hundreds—even thousands—of dollars by shopping for airplane financing as you would any other consumer item. You'll also weigh what you have learned about ownership against the advantages of a leaseback to decide whether that popular method of acquiring a plane is right for you.

Finally, you'll review other methods of owning your plane. You'll weigh the advantages and disadvantages of partnerships. You'll decide whether plane ownership is for you—and your flying budget.

Shopping the Sky

Before you decided to get into flying, you probably noticed only a few airplanes for sale. But as you move deeper into searching for your own wings, you'll discover that nearly every plane is for sale. Narrowing the field is going to be your biggest problem.

Let's take a look at the many sources of new and used general-aviation airplanes.

Authorized dealers   Plane manufacturers are like car makers in that they prefer to have dealers to sell their vehicles for them. Nearly all plane builders have authorized dealerships in major areas where you can inspect both the newest models and the trade-ins. A check of area phone books should net the names and locations of the dealers in major brands such as Piper, Cessna, Beech, Bellanca, and Mooney.

Independent dealers   Many larger airports that cater to the private pilot also house independent airplane dealers who sell used planes, usually on consignment for the owners. They may even carry new planes made by the smaller manufacturers: Maule, Taylorcraft, Rallye, Varga, and Teal. Again the phone book is your best starting point.

Auctions   Growing in popularity, aircraft auctions are a good source of used for-sale planes. You'll see them advertised in the classified sections of major city newspapers and flying publications. They offer good buys for the pilot who knows what to look for and at.

Trade-a-Plane   By far the most popular publication for buying and selling planes is *Trade-a-Plane* of Crossville, Tennessee. Three times a month this advertising newspaper features as many as 100 pages of ads from both

Sunday aviation shoppers inspect some of the planes for sale at a nearby airport in search for the best plane for the budget flier.

businesses and private parties who are buying and selling planes and accessories. It's an excellent source of information on both planes and prices.

Flying publications Other sources of planes for sale and articles about them are *The Aviation Consumer, Flying, Plane & Pilot, Private Pilot, The AOPA Pilot, General Aviation News,* and *Wings.* Most of them can be found in larger libraries or in the pilots' lounges of airports.

Newspapers Most metropolitan newspapers have classified ad sections in which planes for sale are listed under Aviation or Airplanes. Following the classified ads will give you an idea of both what's available in your area and what the prices are.

Bulletin boards Most general-aviation flight offices have bulletin boards for pilots on which notices of planes for sale are posted. Check the ones in your area for a plane of the type and price range you're looking for.

Pilots One of the easiest ways to discover what's for sale at your nearby airport is to ask one of the pilots. Within their close-knit group, pilots have an interest in what's available. They may also venture opinions on a particular plane, its care, and its value. If accepted as opinions rather than fact, that can be helpful to the new pilot who is trying to buy a plane on a budget.

Since ads are expensive, most sellers use a number of widely accepted contractions and acronyms to list their planes' features in the shortest space. Here's an example:

> 1971 CESSNA 150, 2250 TT, 550 SMOH,
> 300 NAV/COM, ELT, XPDR, new annual.
> $7500. Minn. (612) 555-2840.

Translated, it reads: 1971 Cessna model 150 with 2250 hours total time (TT) and 550 hours since major overhaul (SMOH). It features a 300-channel navigation and communication radio (NAV/COM), an emergency locator transmitter (ELT), a transponder (XPDR), and a recent annual inspection, which is required. Minn. is Minnesota, the location of the plane.

Here are other terms you'll run across in ads for budget planes:

AD	Airworthiness directive
ADF	Automatic direction finder
ASI	Airspeed indicator
Cont	Continental engine
DG	Directional gyro
Frank	Franklin engine
FRME	Factory remanufactured engine
GPH	Gallons per hour (rate fuel is used)
IFR	Instrument flight rules, rated for
Lyc	Lycoming engine
OBS	Omni-bearing selector
RB	Rotating beacon
RNAV	Area navigation
SCMOH	Since chrome major overhaul
STOH	Since top overhaul
STOL	Short takeoff and landing
TTAE	Total time, airframe and engine
VFR	Visual flight rules
VOR	VHF omnidirectional range station
VOR/LOC	VOR and localizer

There are others, but the list should help you translate most aviation ads offering the simpler planes of interest to the budget flier.

Check It Out

Once you've narrowed the field down to a handful of planes, you'll want to make a visual inspection to be sure the plane is in good repair. In brief, here are some of the things you'll want to check for.

Fuselage   Examine the outside and as much of the inside of the plane as possible. Check for rust and corrosion, loose rivets, and welds. If the skin is metal, check for serious dents and creases. Inspect the paint for under-lying corrosion and useful life. If the plane is fabric-covered, have a licensed airframe mechanic test the fabric for strength. Also inspect controls for free and easy operation, the instruments for proper operation, and the electrical system for condition and obvious repairs and additions. Examine hydraulic lines and tubing for age and wear.

Mechanics   Open the engine cowling and inspect the unit for the condition of incoming controls and fuel lines. Inspect the carburetor and gaskets. Check the intake and exhaust manifolds for leaks and general condition. Make sure that necessary parts such as the magnetos, starter and generator or alternator, oil sump, and carburetor are attached securely and are in good condition. Then inspect the propeller for cracks and nicks, especially on the leading edge. Make sure the prop is securely mounted. Check the brakes and connecting lines to make sure they are in good condition and are secured.

Logs   The airframe and engine logbooks should show you the inspections and repairs that have been made to the plane. They should also tell you whether there are recurring Airworthiness Directives that will require specific inspections and repairs on a regular basis—an added expense of owning the plane. You can contact the General Aviation District Office in your area to find out what ADs have been issued by the FAA for the plane you're considering.

Any or all of these inspection tasks can be made by a licensed A&P mechanic depending on your own mechanical knowledge and your pocket-book. Having a reputable mechanic inspect a plane that you're seriously considering as a purchase is an investment rather than an expense.

The flight inspection is also important; and if you don't have your license yet, it should be done with the help of an experienced pilot who has flown many types of planes. Since flying begins with a preflight inspection, so should your test flight. Walk around the plane and carefully check all the things you should before you go up on any flight—except even more so.

Inspect control surfaces, leading edges, tires, lights, skin, struts, shocks, windows, tanks, gas level and purity, and anything else that affects your safety in flight.

Once aboard, inspect the interior for comfort and condition. Review the instrument panel to acquaint yourself with the controls, and make sure that instruments are working. Pull out the plane's checklist and run through it slowly. Taxi out for the runup and make a deliberate final check before heading for the runway in use. Let the owner or salesperson help you in the takeoff, and you're soon airborne.

One go-around is not going to tell you what you need to know about the plane's handling, so take as many turns as you need to ensure your satisfaction before you make the final approach. Go through basic maneuvers, shoot a half-dozen touch and go's, and spend some time cruising to give you a chance to ask questions.

Back on the ground, ask to look at the owner's expense records. What were the yearly maintenance costs? What were the fuel costs? What is the plane's typical gallons per hour rate of fuel use? How much is the insurance policy and what does it cover? How much does storage cost at this airport? What is the total yearly cost of owning and operating the plane? How many hours was it actually used? What was the hourly cost of operation?

Then ask for time to consider the plane. When economy is important, a hasty decision can be especially costly and may put a dark cloud over your flight path. Take time to consider all the features and benefits in comparison with the costs. Airplanes that cost $5000 to $30,000 are not purchased like tubes of toothpaste, especially by the budget flier.

Once you've reviewed the purchase of a number of planes, you can make the right decision on the best plane for your needs and your budget. Before we get into financing, let's take a last-minute look at some plane-buying tips that can save you hundreds of dollars and as many hours of frustration:

ADs   To keep you posted on Airworthiness Directives on your plane, you can subscribe to services, such as the Adlog System (Aerotech Publications, Box 528, Old Bridge, NJ 08857), that send copies of ADs on your plane as they are issued by the FAA.

Service difficulty reports   The FAA also keeps on file malfunction or defect reports voluntarily submitted by plane owners and mechanics who have had specific problems with certain aircraft. You can get printouts of such reports through the FAA Aeronautical Center in Oklahoma City, OK 73125, at a nominal charge. They will give you an idea of possible problems you may have with a specific model you're considering.

Budget fliers can justify fitting a better plane such as a new Cessna Skyhawk, into their budget by taking advantage of leaseback or using their wings as a business expense. *(Cessna Aircraft Company)*

Aircraft values   Three used-airplane pricing blue books are used by dealers and leaders: *Aircraft Price Digest, Used Aircraft Price Guide,* and the *ADSA* (Aircraft Dealers Service Association) *Blue Book.* A phone call to your aircraft lender will give you the value of a plane you're considering as listed in one or more of these pricing guides.

Other owners   Once you've found the model you want, talk with owners of the same model in your area about special service requirements and handling characteristics. They may even be familiar with the specific plane you're considering.

Financing Your Wings

You've found the ultimate flying machine for you—or at least an affordable facsimile of it—and you're ready to buy it. Here's how.

First, compare the purchase price asked by the owner with the price that other buyers are paying for the same plane. That's easy enough when you use one or more of the three aircraft blue books just mentioned and check it against asking prices for similar models in your area. Make sure you're not paying too much for your plane.

Next, estimate your down payment. Most lenders will require a minimum of 20 percent down and will furnish the 80 percent balance if you promise to pay it back in "easy monthly payments." Budget fliers make the smallest down payment they can—20 to 25 percent—and save their cash for operating expenses and emergency repairs.

Then estimate your monthly payments with a loan amortization book available in most stationery stores. New planes are usually financed for six to ten years, and used planes are stretched out over four to six years. As an example, a used plane purchased at $8000 might have a down payment of $1600 and 60 equal monthly payments of $142.37 at 12 percent interest.

There are many sources of financing for your aircraft purchase, and you should shop for the one that offers the lowest interest rate, longest loan term, and ease of lending. The best places to try are these:

Your dealer   If you're buying your plane through an authorized dealer, you may be able to finance the purchase through the dealer and the manufacturer's credit plan. Many independent dealers also offer their own in-house financing. Compare their rates and terms with those of other lenders before you decide which plan to use.

Your bank   The bank you normally trade with may or may not have a loan department familiar with aircraft financing. Ask. If it does have, or will make an exception, you may find the easiest terms where you have done most of your business.

Other local banks   Other banks in your area may be able to help you with the specialized financing of an airplane. Ask your own banker to recommend one of them, or call a few banks and ask for the aircraft loan department. Your loan application can be approved much quicker by a local bank than by an out-of-state bank.

Specialty lenders   There are many national lenders that specialize in aviation loans. The best source of their names is the ads that appear in aviation magazines like *Trade-a-Plane, Flying,* and *The AOPA Pilot.* In many cases you can contact them on a toll-free line with your application and have approval in a short time.

The owner   The owner of a low-price-tag plane will sometimes own it free and clear and may consider handling the financing personally. You would make your down payment and monthly payments directly to the seller. If you decide to go that route, have your attorney draw up the papers to ensure that both buyer and seller will be happy with the agreement.

You can learn a great deal about aircraft financing and the lenders themselves by talking with pilots who have financed their planes, which is most of them. Ask your pilot friends whom they would recommend, whom they wouldn't, and why. Also question them about the interest rates and down-payment requirements they faced in financing their own planes.

Your application for aircraft financing is much the same as an application for a loan on a car or any other big-ticket item. The application itself will ask about both you and the plane you wish to buy. The current owner may be able to help you with statistics on the plane model, serial number, equipment, and history. The loan officer will help you fill out the personal credit part if necessary. If you have some of that information ready when you go in for application, you will both save time and make a better impression on your lender. You should have loan approval within a week, and your wings now belong to you—and the bank.

The Leaseback Controversy

As you search for your wings you'll hear a lot, especially from authorized dealers, about leasebacks. Simply, here's how leasebacks work. The dealer sells a plane to you with the promise of leasing it from you in order to rent or lease it to other pilots.

Why would anyone want to do that? The pilot, first of all, to both have a personal plane and take advantage of the tax laws by claiming some of the purchase price and most of the expenses as deductions. The seller—usually a fixed-base operator (FBO) or aircraft dealer—gets to sell a plane and also make additional money by renting the plane to pilots on a per-hour or long-term-lease basis. It can be a sweet arrangement for both buyer and seller.

The emphasis is on *can*. Many factors that enter into the leaseback can make the arrangement a headache. If your FBO has trouble renting or leasing your plane because of low local demand or too many planes already on leaseback, your income from rental will be small and you will still have to meet the payments, the higher insurance premiums, storage charges, and other fixed expenses. If you've purchased a new plane beyond your normal means in the hope of paying for it through leaseback income, you may drain your flying budget quickly and not be able to afford to operate the plane you own.

However, the leaseback plan can give the cautious budget flier a good plane at an affordable price. The best way to make the decision for yourself is to talk with an accountant about your own tax shelter needs and with your FBO about the specifics of renting or leasing and maintaining your plane once it's purchased under a leaseback arrangement. A leaseback can also be successfully worked out with a used airplane. The tax savings are lower, but so is the total purchase price. If the owner-pilot chooses one of the more popular planes such as the Cessna Skyhawk or Skylane, Piper Archer II or Warrior II, or the Beechcraft Sundowner 180, the highest degree of plane use and the greatest advantages from a leaseback are assured.

You'll find more on leasebacks in both Chapter 5, on renting your wings, and Chapter 11, on how to make your plane pay for itself.

Making It Yours

Let's talk about title.

Title is simply the registered right to ownership of a piece of property, in this case, an airplane. The title to a parcel of real estate is recorded in the county in which the real estate is situated. The title to an airplane is registered with the Federal Aviation Administration in the Aeronautical Center in Oklahoma City, Oklahoma. When you purchase your plane, you will fill out an Aircraft Registration Application with the name(s) of the owner(s) of the plane, the N-number of the plane, the make, model, and serial number of the craft, and the signature(s) of the owner(s).

Each plane must also have, somewhere in the craft, a Certificate of Airworthiness, signed by a representative of the FAA, that says the plane is airworthy when maintained within legal limits. Manufactured planes get their certificates as they leave the factory. Builders of home-brewed planes must apply for their own certificates and have their crafts inspected by the FAA before they can fly them legally. When purchasing any plane, make sure that the certificate is available and, preferably, mounted conspicuously in the cockpit.

Another piece of paper you'll want at the time that you buy your plane from a dealer or private party is an Aircraft Bill of Sale (FAA form 8050-2), which is a written agreement from the seller to the buyer to transfer an aircraft and title. If the plane is owned by a partnership and only a share is being transferred, title to that share can be transferred by recording a legal agreement to that effect with the FAA.

Transferring title to and registering an aircraft are usually smooth tasks. However, sometimes the chain of title seems broken and you may not be dealing with the person listed as the registered owner on the plane's Certificate of Aircraft Registration. If so, don't attempt to purchase the plane and reregister title until the current owner has cleared the title of the craft with the FAA.

To ensure that the title to the plane is clear and that there are no hidden lienholders who may be able to sue for title, a title search at the FAA's Aeronautical Center is necessary. Much like the title search done for a piece of real estate, a plane's title search requires a look at documents of registration and modifications. It can be done by your lender through one of the many aircraft title companies near the FAA office in Oklahoma City. The cost is normally $15 to $25.

Forerunner of today's 152, the Cessna 120 is still available in the marketplace for $4000 to $6000. *(Cessna Aircraft Company)*

You can also purchase title insurance that will protect you against unrecorded liens at a cost of about $5 per $1000 of value. Title insurance for a $12,000 plane would cost about $60, which is usually paid by the seller.

Strange Bedfellows

Often the easiest way to buy a plane is by yourself. Your plane is always waiting for you. You don't have to worry about hidden damages caused and unreported by your partner. You don't have to pay someone else's bills. It's simple: one plane, one pilot.

The complexity begins at bill-paying time. You have no one to share the maintenance or storage costs with. All repair costs come out of your flying budget. You cannot buy that new plane you want because you can't afford it all alone.

The controversy of sole ownership versus partnership began over 75 years ago when Wilbur and Orville flipped a coin to see who would fly the plane at Kitty Hawk. It's still a trade-off. You can share the plane and its expenses, or you can have both of them to yourself.

The advantages to individual ownership include exclusive use of your plane, lower aircraft insurance rates, and predictability. The disadvantage, of course, is the higher cost of operation and the lower amount of cash available for the purchase of the best plane for you.

The advantages of partnerships are numerous also. A partnership allows you and a fellow pilot to pool your cash and buy a better plane than you could buy on your own. It also helps you cut operating costs by sharing them with another pilot. Disadvantages include potential conflict over times when

Planes can be found for nearly any application. This Maule STOL plane is equipped with floats for quick takeoffs from lakes and rivers. *(Maule Aircraft Corporation)*

the pilots want to use their plane and the chance that damage caused by another pilot will curtail your flying until repairs are made.

Here are some ideas passed along by high-hour pilots that can help you solve the problem of selecting the right plane and partner to fly with:

Choose your partner as you would a spouse: very carefully. Look for common interests and goals. Match your flying budgets and work together to estimate costs of purchase and operation. Develop a list of minimum features you desire in your plane and match them with your partner's. Come to an agreement before you go shopping for your common plane.

Plan the use of the plane in advance. Try to schedule it so the partners share it equally. If both partners have the weekend off, set aside Saturday for one pilot and Sunday for the other. Split the remainder of the week up appropriately. The best partners are those who work different shifts or have different days off.

Have at least one partnership meeting a month to pay bills, work out scheduling, and discuss upcoming maintenance.

Put your partnership agreement in writing. Answer such questions as how a share can be sold, whether other partners need to approve the sale, what to do in case of a partner who is behind in funds, and how to handle disagreements.

Any partnership is only as good as the ability of its members to work together toward a common goal. That is especially true of the goal of flying on a budget.

Owning your plane, either individually or in common with other budget-conscious fliers, can offer you the freedom of flight at a comparatively low cost. The key is in learning how to buy the right plane for your needs at the right price and owning or leasing it as it best suits your own needs and goals.

Buying your wings can be as much fun as flying them.

5 RENTING YOUR WINGS

Flying is the second most enjoyable activity in the world!

Unfortunately, its cost can deny it to the potential pilots on limited budgets who think they must own the planes they fly. They're wrong. Thousands of new and veteran pilots expand their horizons by letting someone else own the cow while they enjoy the cream. They rent their wings.

Those budget fliers realize that the initial cost of plane ownership—even with a partner—could cut their flying budgets to where they couldn't enjoy the planes they owned. Many also realize that they wouldn't use their planes frequently enough to justify the high fixed costs. It's cheaper for them to rent their wings as they need them.

There are many reasons why fliers rent planes, and not all of them have to do with economy. For some, renting an airplane on an as-needed basis makes the most sense. For others, the wide variety of rental planes is the primary reason for not owning wings. The reasons are as individual as the pilots.

Doing Your Time

Flying skills take time to develop—flying time. Once you earn your private pilot license you may want to go for your instrument rating, which requires 200 hours of flying time and more lessons. An instrument rating will allow you to fly in weather that grounds the VFR pilot. Especially in the northern states, an instrument rating is handy to have.

That is another reason why some pilots rent rather than own their planes. They want to build up their skills on simpler planes than they will eventually buy. They want to keep down the cost of going up.

There are many ways of renting your wings economically. They include straight rental, leasing, and flying clubs.

Straight rental lets you check out a plane, plop down your cash, and enjoy the use of someone else's plane for a certain amount of time, usually counted in hours.

Leasing offers the advantages of nonownership with the lower costs of a long-term commitment. It is of special interest to the business pilot.

Flying clubs offer some of the benefits of both renting and partnerships. A flying club offers a variety of planes, one of which can often be used at a cost lower than that of straight rental.

There are also things you can do as a smart aviation consumer to cut costs within those three methods of renting your wings. You can lower the costs and increase the pleasures of flying on a budget.

Paying the Rent

Renting airplanes is one of the most popular methods of enjoying flight at low cost. In fact, many FBOs are constantly searching for investors who will purchase new planes for leaseback as rentals. They have difficulty keeping up with demand.

Many pilots rent their wings to keep costs down because they don't use a plane often enough to justify the initial, fixed, and variable costs that come with ownership. They fly 50 hours or less per year.

Other pilots own their own planes but use rentals to test models that they are considering for purchase. They may spend their weekends in a Piper Super Cub but have their eyes on, say, a new Beech Sundowner. They decide to rent or even lease the Sundowner to get a long-term look at the next plane they want to invest in. It makes good sense.

Some plane owners periodically rent another plane to be more cost-efficient. A pilot who is completely satisfied with climbing through the sky in a Cessna 120 may want to rent a Mooney Ranger once in a while for the extra speed and seating needed on a family vacation.

Still other pilots prefer to lease rather than rent their planes for both utility and tax considerations. A pilot who leases wings can use a smaller Piper or other plane for reduced business or increased pleasure and a larger Beechcraft or even an amphibious Lake when business is up and needs are greater. In that way the wings are tailored to business and personal needs.

Your choice of a rental plane and the method you use to pay the rental depends on your needs and flying budget. The aviation market is quite flexible, and it can be adapted to the needs of nearly any pilot. You can use the techniques learned in Chapter 3, on choosing your wings, to help you find the right rental and the right plan based on how, how often, when, and why you want to fly.

Shopping for Specials

The price of renting a plane can be as varied as the price of any other consumer item, especially when there are many smaller and independent FBOs whose overhead is less than the municipal airport FBO pays. As a smart budget flier you can take advantage of the disparity in fees by shopping around in your area for low-cost flying time. Here's how:

Choose your equipment   Decide, either through your own observations or those of other pilots whom you trust, what type of plane you prefer to rent. Make your choice wide to cover planes of the same type made by many different manufacturers.

Choose your area   Outline an area on a map which is within an easy driving distance of your home. To reduce the cost of flying, don't consider rental planes outside that area.

Contact rental owners   Call FBOs, new- and used-plane sales offices, leasing offices, and other rental sources. You'll find their locations and phone numbers in the telephone book, in classified ads in metropolitan newspapers, or on airport bulletin boards or get the information from other pilots.

Ask about equipment   Question the rental owner about the make, model, year, and equipment offered in rental planes that you would consider. Make notes.

Ask about price   What is the cost per hour to rent the models offered? Is a deposit required? Is there a discount for longer-term rental?

Ask about minimum time   Does the owner require that the plane must be rented a certain number of hours? Is there a minimum fee? What is the charge if the plane is rented overnight but flown only three hours?

Ask about extras   What does the rental fee include? Gas? Insurance? If so, how much and what type of policy is it? Does it include coverage for passengers? If so, for how many? For personal belongings? Who is the insurance underwriter?

By making notes as you call rental owners, you can compare costs versus value to decide on the best rental plane for you.

As important as which plane you rent and how much you pay for it is the integrity of the person with whom you deal. The ideal rental owner is respected in the local aviation community, keeps the rental planes properly maintained, is easy to reach in case of an emergency, is honest, is helpful, and answers your questions on rentals patiently.

Checking Out Your Rental

Once you've chosen one or two of the best rental buys for you, it's time to go take a look at them in person. The checkout is going to serve two purposes. It will help you decide whether the equipment is what you want to rent and the owner is someone you want to rent from. It will also help the owner look you over carefully. A thorough checkout can save many hours later when you return to rent the same or another plane.

The rental owner's checkout of you will include a look at your pilot license and your logbook. The owner will want to know how many hours you've flown and in what type of equipment, who instructed you, and how long you've been flying and where.

Just as important, the owner will want to discover your attitude toward flying and using other people's planes. That's where some conversation and a check ride come in. The check ride is simply a short flight that allows you to check out the plane you're renting and allows the owner to check out your flying skills before handing you the keys and waving goodbye.

The check ride begins with a ground check of the plane. Go through your preflight inspection as you would with any plane and then ask the owner for the plane's checklist for the start-up and runup. Review the owner's manual for stall speed, best rate of climb, best angle of climb, and best cruise speed. Taxi and take off.

The rental owner will be especially interested in how safely you handle the plane. Don't try any hammerheads or other aerobatic maneuvers. If you're renting the plane to take you on a trip, fly the plane as if you were on that trip. Climb to the proper altitude, trim your controls for level flight, and cruise. Then begin asking questions about the plane's idiosyncrasies; all planes have them. Some planes need a heavier hand on the controls, and others will resist a firm control wheel. You can learn such things with time and experience, or you can ask the owner to share them with you.

Back on the ground you'll have more questions for each other. The owner will want to have your name and phone number, along with your pilot certificate number and rating. You will want information on how to communicate in an emergency. You'll shake hands and be ready to take someone else's $10,000 to $50,000 investment into the sky with mutual confidence.

Renting a plane becomes much easier after that. When you want to rent this or another plane from the owner or FBO, you simply call up and reserve the plane. Its owner knows that you will operate it as you would if it were your own.

Leasing the Sky

Another alternative you have as a budget flier is leasing wings. For many reasons, leasing is more popular with the business flier than with the pleasure flier. First, leasing a plane usually costs less than buying it. Second, leasing offers many tax advantages to the businessperson that the individual cannot take advantage of as readily.

If you're a business flier or will use your wings for both business and pleasure, here's why you should consider leasing:

Less cash is needed   Working capital can be used in other parts of your business rather than in the down payment of your aircraft.

Leases are flexible   You can tailor a lease to your own business needs—with or without gas, insurance, prepaid maintenance, and other expenses. You can also lease a plane for a shorter term than is practical for ownership.

Expenses are deductible   The entire cost of the lease and related operating expenses is a direct operating expense.

Option to buy is possible   You can often include an option to buy your plane after a specified time if you wish to. You can then retain the plane or sometimes sell it at a profit.

Many smart businesspeople use their planes both as transportation for themselves and their employees and as an aid in impressing new and old customers. The cost of taking an important client to lunch at a nearby airport by plane is as deductible as any other business-related use, and it's a whole lot more fun.

In searching for the right lease for you, use the steps suggested for choosing a rental plane with special emphasis on the terms of the lease. You're going to have to live with those terms for from six months to four years or more, so the smart businessperson–pilot will have both an accountant and a lawyer read the fine print in any lease that is entered into.

Join the Club

Flying clubs are becoming increasingly popular, especially those set up by factory-authorized plane dealers for the benefit of new flying students. They offer the benefits of low initial costs, lower rental fees, a variety of planes from which to choose, and the camaraderie of sharing new experiences with people of similar interests. Flying clubs seem to have been designed for the budget flier.

Flying clubs come in all shapes and sizes. They range from clubs with a few members sharing a single plane to 200-member regional clubs with 20 or more planes. Some clubs are developed by dealers for their students and former students, such as the Beech Aero Clubs; other clubs are more independent and offer co-ownership of many types and brands of planes.

Actually, a flying club is just a partnership with more members and a written set of bylaws to govern the members in owning and operating their common equipment. Almost every flying club has a small clubhouse or office at the airport from which the members fly. It can range from a desk in the corner of the flight office to a separate building and hangars featuring a pilots' lounge and a small pilots' store.

Many flying clubs are set up with the aid of a handy booklet produced by the FAA entitled "Forming and Operating a Flying Club" (publication AC 00-25). It's available from your regional General Aviation District Office (GADO) or from the main FAA office in Washington, D.C. It will show you how you can set up your own flying club, choose a plane, write bylaws, and handle minor differences among the members.

Climbing Aboard

The concept of the flying club is that many pilots pool their resources to share ownership and operation of one or more airplanes. Rather than share a down payment, a club member buys into the club and purchases stock in all of the equipment owned by the club. That is done with an initiation fee that can range from $50 to $250 or more. It is a one-time cost.

A club member then pays monthly dues for the continuous operation of the club. Dues can range from $10 to $40 or more a month, and they cover overhead, storage, insurance, payments on the planes, and other fixed flying costs.

Finally, the member pays an additional rental fee for each hour flown. It is usually much less than the normal rental fee from a regular FBO. Also, it

The Cessna 150 is popular on the rental market not only because of its fuel efficiency but also because it can be rented for $12 to $24 per hour including gas and insurance. *(Cessna Aircraft Company)*

includes the variable costs that go up as you do; gas and maintenance reserve are among them.

Whether joining a local flying club is more economical for you than either rental or a normal partnership is an individual decision based on many factors such as the availability, quality, and cost of local flying clubs. Let's take a look at the cost of membership in a typical flying club operated by a dealership:

	Initiation Fee, $	Monthly Dues, $
Student	65	15
Intermediate	125	20
Advanced	185	20

Rental fee on a new Cessna 152 would be an additional $17 per hour solo. By amortizing the intermediate initiation fee over two years and estimating five hours of flying a month, you find that your cost would be $110.21 per month or about $22.04 per hour. You can compare that figure with the rental fee for the same plane from a fixed-base operator to make the economic decision of whether flying club membership is worthwhile to you.

In shopping for a flying club, remember that there are two basic types of club and that one type may have advantages over the other for you:

Shareholder flying clubs In this, the oldest and most common type of club, members purchase a share in the club with their initiation fee and actually own a percentage of what the club jointly owns. The club takes care of maintenance and other fixed costs out of monthly dues. You pay variable costs with your hourly rental fee.

Leaseback flying club   A newer idea in flying on a budget, the leaseback club has a minimal initiation fee because the planes are actually owned by investors who have purchased them to lease them back to the club. Monthly dues are usually about the same as in a shareholder club. The hourly rental fee may be slightly higher, but it is still less than the normal rental fee.

The size of the club you join should also be a factor in your decision because size dictates both cost and availability of planes.

A small flying club allows you to know and share with a small group of pilots. The club may only own one or two planes, but each pilot, on stepping into the common bird, has the comforting assurance that the plane's other users are as concerned about safety and courtesy as he or she is.

Of course, a smaller club may also mean fewer members to share the costs of a major overhaul or other catastrophe.

A large flying club with dozens or even hundreds of members offers you the buying power of many in selecting a wide variety of aircraft. A larger club may offer you the use of everything from a Maule STOL to a Beechcraft Bonanza, plus a full-time instructor at a reduced rate.

The disadvantage to a large flying club can be the loss of first-hand friendship with every member. You may not even know all the members by name. Also, large clubs can have a greater faction of dissent.

The decision is a personal one. Of course, it may be made for you by local availability of clubs and their costs of membership. You may want to join a smaller club but discover that a larger flying club fits more of your flying needs.

To help you decide whether joining a flying club is best for you and, if so, which one to join, here's a checklist you can use:

What flying clubs are there in the area?
Where do they operate from?
Are they independent or dealer-affiliated?
How many members belong to the club?
What type of club is it? Shareholder or leaseback?
What type of equipment does it own? How many planes?
Who handles maintenance and how often?
Does it do preventive maintenance regularly?
Do I know any of the members? What do they think of their club?

What is the initiation fee?
What are the monthly membership dues?
What are the rental fees?
Do rental fees include gas and insurance?
If I decide to leave the club, may I sell my share? How?
What will this club require of me?
Are members new, intermediate, or advanced pilots?
How long has this club been in operation?
Who are the club's officers?
Are memberships currently available?

The field of flying clubs can be narrowed down easily by a study of your own flying needs and budget against the offerings of local flying clubs.

Getting the Most from Your Club

Once you've joined a flying club, there are many things you can do as a smart aviation consumer to make sure you get your money's worth. For one thing, you can get low-cost instruction.

Many flying clubs are set up by individuals or plane dealers to make it easier for people to earn their wings on a budget. To reduce the cost of renting a plane and instructor, the club often makes special pricing considerations to student pilots. Initiation fees or sign-up charges are often lower for students because of the lower cost of trainer planes and a need for new members, and so are monthly dues.

Instruction also may be lower in cost to members of a flying club. Larger clubs often have full-time administrators who also act as instructors for members who are going for their private pilot licenses or instrument ratings. Also, some members of the club may be certified flight instructors who will give instruction to other members at a lower fee than a regular school would.

The savings of instruction through membership in a flying club can be as much as a quarter to one-third of the cost through a regular flying school. Once you have your license, membership in a flying club can continue to save you money and help you fly on a budget.

Most clubs have a location where you can reserve a plane in advance by simply signing up on a schedule. You can reserve it for business or pleasure.

A high percentage of club flying is around the windsock. That is, most members log many hours in the landing pattern and within a few miles of the airport practicing for a test or to keep their skills current.

A smaller percentage is cross-country flying of a hundred miles or more. Some members will use their club's plane to take the family on a flying vacation; others will use it to conduct business in other cities. In fact, many smart businesspeople write off the cost of their club initiation, dues, and rental fee by using it as a business expense.

Still other members are using club equipment to upgrade their licenses. They take advantage of the reduced rental costs to build up flying hours and practice for their instrument ratings.

Membership in a local flying club can be one of the most inexpensive ways for a budget flier to enjoy the sky without having to buy it. If you don't mind sharing your wings with pilots with similar interests, you should look into joining a flying club. A club can mean many extra hours of flying fun for the same number of dollars.

Cutting Down on the Rent

For the new pilot, renting your wings—through an FBO, a lease, or a flying club—can be the most economical way to get airborne on a budget. To help you cut costs even further, here are a few ways in which other cost-conscious pilots reduce flying costs:

Take a friend Sky "carpooling" can be fun and can be economical as well. Whether you're heading somewhere in particular or just out for a stroll in the sky, sharing expenses with a passenger can help you reduce costs. Of course, if you charge a passenger to fly with you, you are a commercial pilot and must have a higher license and submit to more stringent maintenance and inspection of your plane. But you can accept a free-will gift of cash to help defray the cost of avgas and other expenses.

Buy a block of time Sometimes you can find a pilot who will sell you a specific amount of time with his or her plane at a reduced rate for cash in advance. You may be able to buy 50 hours of time provided only that you give the owner first opportunity to use the plane. Beyond that, the 50 hours are yours when you need them. The savings can be 25 percent or more over rental from an FBO. But first make sure that the owner's insurance policy will cover your business arrangement.

Ask an accountant A half hour of a good tax accountant's time can save you enough for many extra hours of flying time. Ask about using your rental plane for both business and pleasure. Ask about the tax savings you can earn through joining a flying club over straight rental. Ask about leases and whether it is in your best interest to lease your wings. More about the tax advantages of plane ownership, leasing, and renting in Chapter 11.

Talk with other pilots The best source of current information about local rental fees, leases and leasebacks, and flying clubs is the pilots' lounge at your local general aviation airport. Remember that, by nature, pilots are an independent breed and are often highly opinionated. Listen to the facts and make up your own mind.

Earn extra cash There are many ways in which you can cut the cost of rentals by supplementing your budget with income from flying-related business ventures. Chapter 11 offers the most popular and most productive of such opportunities.

Renting your wings can be the way to go. It can offer you the joy of flight without the problems of ownership. Renting, leasing, and flying clubs seem tailor-made for the budget flier—because they are.

6 BUILDING YOUR WINGS

You say that you have your heart set on a new plane, my friend, but you don't think you can afford it?

You feel that purchasing a plane—new or used—is beyond the elasticity of your budget?

You tell me you don't want to rent and fly an unfamiliar set of wings?

You say you can't see paying a mechanic with a fancy piece of paper on the wall $25 or more per hour to change spark plugs?

You insist that you have more time than money and you'd rather do it yourself?

Tell you what I'm going to do. Step this way, my friend, and I'll introduce you to the mystical, magical, money-saving world of building your own flying machine. This here's the Right Brothers School of Do-It-Yourself Aviation.

I'm serious. Building your own wings is fast growing in popularity as a way to get airborne on a budget. In fact, it's estimated that there are now 10,000 amateur-built planes flying and another 20,000 under construction.

There are many ways in which you can build your own wings on a budget. The first, of course, is to design your own flying machine and construct it within the confines of a garage or hangar. If you know something about aerodynamics and are willing to learn more, you can design your wings economically.

Even more popular are the dozens of airplane kits available on today's market for building everything from a motorized hang glider to a four-set luxury plane. The craft can be built in a week to a year or more depending on your own schedule and what you want for a finished product.

Other smart fliers are rebuilding wrecked planes they buy at salvage prices. With a little knowledge, and/or the help of a friendly A&P mechanic, you can turn someone else's mistake into your low-cost joy.

Or, if you really enjoy the romance of flying, you can rebuild an antique or classic plane that will give you basic flying pleasure at a lower cost. You can own a piece of aviation history and convert your spare time into hundreds of hours of recreation by rebuilding an older plane.

Why do people want to build their own airplanes? For many reasons. The first and often the foremost is cost. More than half the cost of most products we buy today—including airplanes—is for labor. Fliers who are willing to do much of the labor in building or rebuilding planes can often save the thousands of dollars they would indirectly pay others to build them. Actually, they are paying themselves. With the savings, they can either buy better planes or spend more time in the sky.

Amateur plane builders also earn a knowledge of their craft that few pilots of factory-built planes ever have. They build or rebuild their planes piece by piece until they know every strength and weakness. And those who build more than half of their craft have planes that qualify for the amateur-built category. Owners of such planes can do all their own maintenance work without the help of costly licensed mechanics. The home-brewed plane builder can save many hundreds of dollars by legally being the mechanic.

Another benefit of building your own wings is the additional pleasure you get from flying a plane you built or rebuilt. It's a source of added enjoyment. Your wings were chosen and built for a specific purpose, unlike the all-things-to-all-people planes that must taxi out of commercial aircraft factories.

Finally, the amateur plane builder earns the respect of other pilots who realize that the spirit of aviation is carried by those who build their own planes or rebuild the classics of the past.

Sketching Your Wings

If you have a background in aerodynamics, you may decide to design and build your own airplane. It's no easy task, but there are many places you can go for help. The first thing to do is join the Experimental Aircraft Association (EAA) and subscribe to its *Sport Aviation* magazine, where you'll read about thousands of others who are building planes from scratch, from plans, or from kits. The EAA offers numerous services to the plane builder to help with both design and actual construction of the craft.

Books that will help you with your project are also available. They include *Elements of Sport Airplane Design for the Homebuilder, Build Your*

Own Airplane by James Bede, *Engines for Homebuilt Aircraft* by Christy and Erickson, *Restoration of Antique and Classic Planes* by Don Dwiggins, *Guide to Homebuilts* by Peter Bowers, *Light Airplane Design* by L. Pazmany, *Custom Light Planes: A Design Guide* by William Welch, and the highly recommended *Design for Flying* by David Thurston. All these titles are available through Steven's Company and others who advertise regularly in the major aviation magazines.

I'll give addresses of major organizations in and suppliers to the homebuilt and restoration field at the end of this chapter.

If you don't feel you have the knowledge to build a plane from your own or someone else's plans, you can make your wings from one of the dozens of airplane kits available today. The kits range from simple 30-hour projects to complex 3000-hour jobs that require many skills.

The EAA and the major aviation magazines are, again, the best source of information on what kits are available. Many kits have been around for 20 years or more; others are newly introduced or newly retired.

The way to choose a kit is to look at these four elements of building aircraft kits and compare your own needs and capabilities with the features of the many kits available:

Your flying needs Are you looking for the superior aerobatic qualities of the Pitts S-1, the simplicity and low cost of the Easy Riser, or the fundamental nostalgia of the Pietenpol Air Camper? There's a kit that fits every reason for wanting one.

Your budget Kits range in price from under $2000 to over $20,000 depending on the complexity of the finished product. Comparison shopping is important here.

Your knowledge How many building skills do you know or can pick up? Welding? Fabric covering? Engine assembly? How much can you do and how much will you need help with?

Your time How much spare time do you have to spend in the garage or hangar building your plane from your kit? How understanding will your family be?

You will also want to consider Uncle Sam's interest in your project. The FAA must inspect your home-brewed airplane both during construction and before the first flight in order to issue you an airworthiness certificate.

To earn your airworthiness certificate, you are expected to conform to acceptable construction methods and use acceptable materials. Before the box is closed off and the airframe is covered, the FAA must inspect the

Many do-it-yourself airplane kits are available in the aviation marketplace. They include this Wag-Aero CUBy—a near-twin of the Piper Cub—for less than $6000. *(Wag-Aero, Incorporated)*

materials and workmanship to make sure your plane is both legal and safe.

The home-built plane is restricted to flying within 25 miles of your home field for the first 50 hours of air time (75 hours if powered by a nonaircraft engine) to make sure it is airworthy.

As mentioned earlier, the advantage of the amateur-built over the factory-built plane is that the owner can be the mechanic and sign off the required annual inspections on a home-built craft. That is a major reason why many aviators decide to build their planes rather than buy planes prebuilt. With the money they save by being their own mechanics, they can buy avgas.

Planning Your Project

How much time and money is your home-built plane going to cost? Where are you going to build it? When can you schedule time for your newfound hobby?

The answer to those questions are as individual as the planes and the pilots who build them. To help you make your own decision, here are typical prices and construction times for a few popular plane kits:

The Wag-Aero CUBy (a reproduction of the old Piper Cub in kit form) currently costs under $6000 as a kit and takes an estimated 1200 hours to complete.

The Pitts S-1 aerobatic plane can be purchased as a plan for $150, as a kit for under $13,000, or complete and ready to fly for about $33,000.

The Redfern Fokker Triplane, a replica of the Red Baron's flying machine, is available in plan form for about $50. You supply the parts and labor.

The Weedhopper is a small one-person flying machine in kit form priced under $3000 and boasting a 40-hour construction time.

The typical full-sized two-seater plane will take about 1200 hours to build from a kit and about twice as many hours from a plan. That's two hours a day, six days a week for two years to complete the kit and four years to finish your plane from plans. Many pilots don't want to wait that long to have their own wings; but if you are both patient and productive, you can build your own flying machine in your garage or hangar and save thousands of dollars over production models.

Rebuilding Your Wings

You can also save money on your bird by either rebuilding a wrecked plane or restoring an antique or classic plane. This is where the aviation consumer who knows what to look for, knows how much time and money restoring the plane will cost, and knows where to find a good buy can really save cash.

There are many good sources of rebuildable wrecks that can be restored for less money than buying a comparable model in good condition:

Local airports Check the bulletin boards of local general aviation airports for salvageable wrecks for sale and check open hangars. The FBO may know of some; find out by asking. If you aren't one yourself, you'll want a licensed mechanic along to estimate the damage and the cost to repair the bird.

A&Ps Your mechanic may have a wrecked plane or two for sale. Be careful, however. If there were a great opportunity for a profit from fixing the plane up, the mechanic would probably take on the job. But if you know the mechanic well and can get a qualified second opinion, you may find a good buy in the corner of the shop.

Trade-a-Plane Again, the best source for published plane ads is *Trade-a-Plane*. It features an entire column of "Rebuildable Aircraft, Parts, Services, Etc."

Insurance companies Once in a while an insurance company "totals" a plane even though it's easily rebuildable. Call a few companies to see if there are such birds in your area they are willing to sell.

Rebuilding firms At larger airports you'll often find one or more businesses that specialize in buying wrecked planes and restoring them to

Many budget fliers invest in their hobby by rebuilding or restoring older planes such as this Cessna 140. They not only offer inexpensive flying fun, but also climb in value each year. *(Cessna Aircraft Corporation)*

flying condition. In many cases you can furnish the dough and they will furnish the bird and parts and labor to rebuild it for less than a comparable unwrecked unit would cost.

What should you look for in a damaged plane? Well, that all depends on how much you have to invest in the rebuilding and how many headaches you want to trade for dollars saved. Some so-called wrecked aircraft simply have wing-tip damage due to taxiing negligence. Others may suffer from bad cases of hangar rash caused by other planes bumping into them too many times. Most wrecked planes, though, are the result of accidents. The most common damages include:

Wing damage
Strut damage
Landing gear collapse
Propeller damage
Tail damage

Most of these can be handled by a qualified A&P mechanic at a minimal cost compared to replacing the plane. Major damage, of course, should be considered carefully because, once the repairs are started, it's difficult to recoup your investment until the rebuilding project is completed. A major rebuilding project should be started only if the plane has a wide enough gap between salvage cost and value at completion to meet expenses and return a small profit for headaches.

Restoring a Classic

For historical, aesthetic, and investment reasons, many pilots are restoring old planes for fun and profit. Over 5000 aviators and enthusiasts belong to the Antique Airplane Association (AAA), which boasts a museum and its own airfield. The Association is host to a yearly fly-in for hundreds of old planes.

The AAA puts all antique and classic planes into one of five categories listed in the accompanying table. Those aren't dusty museum pieces, either. Most member planes fly regularly. They range from Porterfield-Turners to Luscombes to Stearman biplanes and Beechcraft Staggerwings. They are flying monuments to aviation history.

Pioneer	Aircraft built prior to World War I
World War I	Planes built between 1914 and 1918
Antique	Airplanes built between 1919 and 1936
Classic	Planes built from 1937 through World War II
Neoclassic	Airplanes built since World War II and no longer in production

They're also a lot of fun to rebuild and fly. In addition to the AAA there are dozens of specific clubs that specialize in helping members restore certain antique or classic planes. They include the Ercoupe Owners Club, American Navion Society, International Cessna 170 Association, the Fairchild Club, and World War I Aeroplanes. There are many others. Their collective aim is to "keep 'em flying!"

If you decide you want to rebuild one of the old planes, the best place to look is to one of the clubs interested in that kind of plane. Many antique and classic aircraft are offered for sale through AAA publications and newsletters put out by specific clubs. When buying a plane you're unfamiliar with— from a Waco to a Skyhawk—have someone who is knowledgeable about the type look the craft over before you buy it. Parts for some older planes are no longer available and must be machined at a greater expense. Budget flying means restoring a plane the parts for which are easy to find.

Another source of old aircraft are publications for the general aviator such as *Flying, Private Pilot, Plane & Pilot, Trade-a-Plane, AOPA Pilot,* and *Air Progress.* Planes, parts, and services are available through individuals and small companies that cater to the growing antique airplane field.

Two of the largest parts houses for restorers are Wag-Aero and Univair. Univair manufactures thousands of parts for out-of-production aircraft like the Stinson, Swift, Ercoupe, Forney, Alon, and Mooney M-10. They also handle parts for Aeroncas, the Bellanca Citabria, many Cessnas and Pipers, the Luscombe, and Taylorcraft. Wag-Aero says it has the world's largest inventory of replacement parts and accessories from parts lists to wing sections. Both Wag-Aero and Univair are prime sources for the rebuilder and restorer.

Many of the classic and neoclassic planes offer the budget flier special opportunities to cut costs while enjoying flying. A number of planes of the era were designed as basic economy models; and as long as parts are readily available, they will remain worthwhile investments for the cost-conscious flier. Some can be purchased for less than $5000 with the assurance that values will only go up.

Whether you decide to build a plane from your own design, package plans, or a kit, rebuild a wrecked or damaged plane, or restore an antique or classic, you can both lower your flying costs and increase your flying fun by applying the philosophy of the budget flier: "I'd rather do it myself!"

Organizations for the Budget Builder

Experimental Aircraft Association
Box 229
Hales Corners, WI 53130

Antique Airplane Association
P.O. Box H
Ottumwa, IA 52501

Aircraft Owners and Pilots Association
Air Rights Building
7315 Wisconsin Avenue
Bethesda, MD 20014

American Navion Society
P.O. Box 1175
Banning, CA 92220

Cessna 190–195 Owners Association
c/o Thomas Pappas
P.O. Box 952
Sioux Falls, SD 57101

Cessna 120–140 Association
Box 92
Richardson, TX 75080

Ercoupe Owners Club
3557 Roxboro Road
Durham, NC 27704

International Cessna 170 Association
29010 E. Highway 160
Durango, CO 81301

International Mooney Society
2202 Oakhill
San Antonio, TX 78238

Soaring Society of America
Box 66071
Los Angeles, CA 90066

Swift Association
Box 644
Athens, TN 37303

Taildragger Pilots Association
3039 Kingsgate
Memphis, TN 38118

Suppliers for the Budget Builder

Aircraft Spruce and Specialty Company
Box 424
Fullerton, CA 92632

Aircraft Tool Supply Company
5738 N. F-41
Oscoda, MI 48750

Airtex Products, Inc. (aircraft interiors and exteriors)
Lower Morrisville Road
Fallsington, PA 19054

All Aircraft Parts
16673 Roscoe Blvd.
Sepulveda, CA 91343

Century Instrument Corporation
440 Southeast Blvd.
Wichita, KS 57210

Cooper Aviation Supply Company
2149 E. Pratt Blvd.
Elk Grove Village, IL 60007

Gibson Aviation (engine rebuilding)
P.O. Box 880
El Reno, OK 73036

J & M Aircraft Supply
P.O. Box 7586
Shreveport, LA 71107

Sporty's Pilot Shop
Clermont County Airport
Batavia, OH 45103

Univair Aircraft Corporation
Rt. 3, Box 59
Aurora, CO 80011

U.S. Industrial Tool & Supply Company
13541 Auburn
Detroit, MI 48223

Wag-Aero, Inc.
Box 181
Lyons, WI 53148

Plans and Kits for the Budget Builder

Acro-Sport, Inc. (EAA Acro-Sport Biplane)
P.O. Box 462
Hales Corners, WI 53130

Birdman Aircraft (Birdman ultralight aircraft)
480 Midway
Daytona Beach, FL 32014

Peter M. Bowers (Fly Baby foldable-wing plane)
10458 16th Avenue, S.
Seattle, WA 98168

Evans Aircraft (Volkswagen-powered Volksplane)
P.O. Box 744
La Jolla, CA 92038

Pitts Aerobatics (Plans, kits, and completed planes)
P.O. Box 547
Afton, WY 83110

Redfern & Sons, Inc. (Fokker Triplane replica)
Route 1
Athol, ID 83801

Rutan Aircraft Factory (Canard-style VariEze plane)
Building 13
Mojave Airport
Mojave, CA 93501

Sequoia Aircraft Corporation (S-300, S302, and Falco F.&L high-performance aircraft plans and kits)
900 W. Franklin Street
Richmond, VA 23220

Sky Sports (HA-2M Sportster ultralight aircraft)
Box 507
Ellington, CT 06029

Mrs. F. W. Smith (Pioneering home-built biplane)
7502 Sunny Hill Dr.
Norco, CA 91760

Molt Taylor (Coot amphibious and Mini-Imp planes)
Box 1171
Longview, WA 98632

Ultralight Flying Machines (Easy Riser ultralight)
Box 59
Cupertino, CA 95014

Wag-Aero, Inc. (CUBy, Wag-a-Bond aircraft kits)
P.O. Box 181
Lyons, WI 53148

War Aircraft Replicas (Focke-Wulf 190, Vought Corsair replica plane plans)
348 S. 8th Street
Santa Paula, CA 93060

Weedhopper of Utah (Weedhopper ultralight kit)
Box 2253
Ogden, UT 84404

7 BRINGING DOWN THE COST OF GOING UP

Flying can be a very expensive hobby!

Most of the maintenance required on the average plane is costly. It must be done by a licensed A&P mechanic, whose services will cost you $25 or more per hour.

Aviation insurance is more costly than automobile insurance—especially for the new pilot—because of higher liability requirements.

The fuel costs of flying are rising along with those of other modes of transportation, and the lack of competition eliminates the chances of a "gas war." Avgas is never going to be cheap.

Storage also is more expensive as industrial sites and subdivisions threaten to consume smaller airports and thus drive up taxes, property values, and the cost of storing your bird.

However, there are many things that the smart budget flier can do to lower the costs of flying and increase the amount of flying time per dollar. It's a matter of being a smart aviation consumer.

Maintenance is often the largest expense for the average pilot. I'll devote all of Chapter 8 to showing you how you can legally reduce the cost of maintaining your plane.

Insurance is usually the next largest expense for a flier; it costs as much as $1000 per year. The keys to lowering your aviation insurance costs include understanding what the insurance does and doesn't cover and how much and what kind of insurance you need.

The cost of fuel is difficult to change, but the smart budget flier can use fuel-efficient flying techniques to lower fuel consumption and improve the miles per gallon rating.

You can also save money on plane storage by shopping around, deciding on the best storage for your plane, getting the best terms, and utilizing your storage for cost efficiency.

You don't have to take rising flying costs on the chin—or in the wallet. You can fight back by learning how to cut operating costs and spend more money doing what you want to be doing: using your wings.

Insuring Your Wings

Insurance is a good thing. It takes the risks and worry out of accidentally becoming involved in a lawsuit that can quickly erase all your assets. It gives you a partner who will help you pay for damages to personal property over an agreed-upon amount.

Having too much of a good thing is what makes your insurance premium and operating costs rise beyond necessity. Buying more insurance than you need for your craft is a waste of good flying money.

So why do you even need insurance?

Two reasons: to protect yourself and to protect others. You want to protect yourself from major losses that can occur in operating any vehicle, especially one that travels through the air at over 100 miles per hour. You also want to protect others from expenses they can incur as a result of damages caused by your plane. It's not enjoyable to think about; but accidents can happen, and the high cost of medicine today can mount up rapidly in the event of even a minor accident. The smart aviation consumer buys insurance to cover the risks of flying.

That's what aviation insurance is designed to do: accept some of the risk of potential loss in exchange for some of your cash. It protects you and others against losses caused by or resulting to your airplane.

There are two types of aviation insurance: hull insurance and liability insurance. For most planes and pilots, hull insurance is the most expensive kind. It covers damage to your plane from both accidents and such acts of God as wind and hail. Coverage includes damage to anything in or on your plane. The cost of hull insurance is based on the value of your wings. More expensive wings require more expensive hull insurance.

To lower claims and risks, hull insurance has a deductible. You agree to pay for the first $100, $500, or $1000 of damage, and your insurance company agrees to pay the rest. The higher the deductible, the lower the premium.

If you finance your wings through a lender, you will probably be required to carry hull insurance on the craft in at least the amount of your loan. Your lender wants to make sure that, in the event the plane is totally destroyed, at least the money loaned you is secured.

If you aren't financing your plane, you don't have to purchase hull insurance. It's up to you. Your decision will depend upon the cost with various deductibles, the value of your plane, and whether you could financially bear damage to most or all of your plane.

Liability insurance covers losses you might incur through injury or damage to other people or their property. How much liability insurance you get and how much premium you pay for it depends on how much of your risk you want someone else to accept.

Liability insurance covers claims with limits on:

Each person
Property damage
Each occurrence

A typical liability policy for the average pilot would cover $50,000 in claims from each person, $300,000 in property damage claims, and a total claim from the occurrence of $300,000. Of course, the claims are made by other people; you should have your own life and health insurance policies to cover yourself. Your agent can help minimize wasteful duplication of coverage.

When you are deciding the right policy for your flying needs, consider both your own monetary worth and the capabilities of your airplane. How much financial protection do you need to cover the assets you've built up in life? There's also a limit to how much damage you feel your plane can cause. The amount depends greatly on the speed and seating of your craft. If you're flying a two-seater, your liability is much lower than if you had a four-passenger plane, and your policy limits don't need to be as high.

Buying aviation insurance is like buying any other necessary consumer item: you shop value and service. You want the greatest value for your dollar, and you want service from an agent who will help you in both buying and using your insurance.

There are three ways you can buy your aviation insurance:

Underwriter The person who actually writes the policy and takes the risks is the underwriter. To save the costs of intermediaries, many pilots work directly with aviation insurance underwriters who advertise in the major flying magazines.

Broker Other pilots prefer intermediaries who are knowledgeable in aviation insurance to choose the best policies as well as the best underwriters.

Agent   An insurance agent covers a broader field, often handling many types of insurance from many insurance companies and underwriters. An agent may handle both your aviation and your automobile insurance.

Which way you choose to handle your insurance needs depends on many factors; but if your prime consideration is cost and value, you will usually save more money by going directly to the source and contracting your insurance through an aviation insurance underwriter.

Cutting Insurance Costs

Now that you have a good idea of what aviation insurance is and where to get it, let's see some of the ways in which you, as a budget flier, can lower your insurance costs and get more for your money.

The best way to keep insurance costs down is by owning an economical plane. Sure, hull insurance on an inexpensive plane is much lower than a twin-engine superduper sky-streaker. The monetary value is less, so the replacement cost is less—which makes insurance to replace the plane cost less.

As mentioned a moment ago, seating also affects your liability insurance rates. A two-seat plane just can't cause as much personal and property damage as a six-seat plane.

A more economical airplane is also safer—and thus less expensive to insure—because its top speed is lower than that of a big bird. The chances of injury and damage are increased as the speed at which you're traveling increases.

The point is that a simpler and less expensive airplane can save you money not only when you buy and fly it but also when you insure it.

The type of flying you do has a relation to the cost of your aviation insurance. If you spend your Sundays racing around pylons, your insurance premium is going to be much greater than it would be if you spent that time inspecting the countryside in your Piper Cub.

Insurance rates for business flying and pleasure flying are very similar, so you shouldn't see much of a difference in your premium if your flying time moves from one category to another. If your plane moves from pleasure to rental or instruction, though, expect a jump in costs because of the added risks.

Insurance will cost more if you are the member of a flying partnership. Plane use is greater. You may be sharing your wings with a low-time pilot who will bring the rates up for everyone. The benefit is that, through a sharing of the cost of the more expensive policy, your insurance costs will probably be lower in a partnership than they would be if you were to buy a policy just for yourself. That's another point in favor of partnerships.

If you plane is on a leaseback through an FBO, expect a higher insurance premium that is due to increased risk. Again, though, you are hopefully sharing the premium cost with the pilots who rent or lease your plane.

Your experience as a pilot will affect your aviation insurance premium. It stands to reason that, as your flying skills and time increase, you are less of a risk to your insurance company. You simply make fewer mistakes. So as you earn your commercial and instrument ratings and fly more hours, your premium will decrease. Be sure to update your insurance company on any new ratings you earn as soon as possible to take advantage of reduced premiums and more flying time for the same number of dollars.

One last point to remember about aviation insurance is that it is like any other business in that it must depend on consumers to survive. In most cases, aviation insurance companies will develop and suggest many types of coverage for every conceivable risk—often more coverage than you really need. But competition is such that a comparison shopper who understands both what is really needed and what is available can reduce the second-highest cost of going up: aviation insurance.

How to Fuel Your Flying Budget

The days of the 29.9-a-gallon-and-a-free-dish gasoline are gone forever. All petroleum products are becoming more and more expensive as oil-producing nations play international poker with their resources.

Even so, in many ways aviation gas is a bargain. For as little as three dollars in fuel an economy-model plane can take you to the sky for an hour's visit. You can climb above the cities and glide across the countryside. Think what a person of a hundred years ago would have paid to do what you can now do for the price of a hamburger special.

Most general aviation aircraft use one of three types of aviation fuel (avgas):

80/87 octane (red in color)
100/130 octane (green in color)
100/130 octane low-lead (blue in color)

In general, most of the older low-compression plane engines use the 80/87 octane fuel (80 is the lean-mixture rating and 87 is the rich-mixture rating), which is becoming more scarce. Newer planes with higher horsepower use the 100/130 high-lead gas. As an alternative for both the older and newer engines, 100/130 low-lead has been introduced. It has about half the lead of the high-lead type but four times the lead in 80/87.

Those facts should be remembered when you consider both the plane you buy and the field you tie down at. Can you get the gas you need at the field? Converting an 80-octane engine to a 100 high-lead engine means a new engine. Changing it over to a 100 low-lead usually means an additive to reduce spark plug fouling and a few minor adjustments. Talk with a mechanic to make sure you can use the new low-lead before you buy that 80-octane flying machine.

Most small planes have two tanks aboard to store fuel. The tanks are usually located in the wings near the cockpit. How they feed gas to the engine depends on whether the wings and wing tanks are above or below the engine. High-wing planes use a gravity-feed system and low-wing planes a pump- feed system to get fuel to the engine.

You can also purchase optional wing-tip fuel tanks for most planes to extend the flying range of your bird. Depending on how you use them, wing-tip tanks can increase or reduce your plane's fuel efficiency. Here's why: You best fuel economy will be found at higher altitudes, where the air is thinner and the plane needs less lift and thrust to fly. About 8000 feet is often suggested. If you're flying a long trip, wing-tip tanks will extend your range and allow you to maintain optimum altitude longer without refueling. The drawback is that more fuel means more weight. A gallon of gas weighs about six pounds. So for shorter trips, wing-tip tanks should be nearly empty to reduce weight and increase fuel efficiency.

Cut the Fuelishness

There are many things you as the pilot can do to reduce the amount of fuel it takes to get there and back. And it all begins with the preflight.

Once you've started your plane's engines, the Hobbs meter on it begins ticking away the time in tenths of an hour. Whether you are operating your own plane or a rented or leased plane, the Hobbs meter is taking money out of your pocket. The key to preflight fuel economy is to not turn the engine over until you're ready and, once you do turn it over, to use the least amount of time for your preflight check.

That doesn't mean that you should switch the engine on and then rush blindly through a cursory checklist. Rather, it means that you should be familiar with your cockpit and memorize the procedure for a quick but thorough preflight inspection. Memorize and understand each step of your plane's preflight checklist. Know where every instrument and control is and know what it's supposed to read or do. Many experienced pilots save even more time by running through part of the checklist as they taxi toward the departure runway.

Another way to save on fuel is to reduce the weight you carry around. It has nothing to do with a diet. It's a half-minute check of your plane's baggage area to make sure you don't have more weight aboard than you really need. Take the necessary parts, kits, and tools for safe flying, but leave everything unnecessary to your plane's flying or your own needs behind. An extra hundred pounds can drain away that extra gas reserve you may need later.

While you're checking the load, make sure it is near the plane's center of gravity. Your owner's manual will tell you where that point is on your plane. An unbalanced plane can take more fuel to do the same job. It can also cause control problems. Fly by the book—the owner's manual—and you can save gas money with an efficient preflight inspection.

A smart budget flier can also save fuel in the takeoff and climb segments of flying. As student pilots we are taught to shove the throttle to the wall in takeoff and get the bird flying as soon as possible. That's good advice for the student. But as you gain experience with both your plane and your home field you'll be able to lift off quickly and safely in little more than half the runway's length. Then it's time to save fuel by reducing your takeoff throttle setting and using more of the runway. A little practice will tell you the optimum throttle setting for getting you off the ground before you run out of pavement. Budget flying means flying both safely and economically.

Student pilots also learn four climb speeds: normal climb speed, cruising climb speed, best rate-of-climb speed, and best angle-of-climb speed. The most fuel-efficient is, in most cases, the normal climb speed. It is chosen as the optimum speed at which your plane can gain both altitude and ground speed. Check your owner's manual for the normal climb speed that will give you the best fuel economy.

Once you're up in the sky, you can improve fuel efficiency by choosing the right altitude. As mentioned before, higher altitude means less drag and gravity to slow your plane down. Since you'll need oxygen if you fly much above 10,000 feet, the most efficient altitude for cross-country flying is between 6000 and 8000 feet. Climbing to 6000 feet above ground level to (AGL) at 75 miles an hour and 500 feet per minute, for example, will take about 12 minutes, or about 15 miles. If you're flying only 30 miles, you'll have to begin your descent as soon as you reach your cruise altitude. Not very efficient. For shorter runs, use a cruise altitude that will allow you to operate more often at the more efficient cruise speed rather than climb speed.

Ideally, you've planned your trip to do most of your flying at cruise speed and altitude for greatest efficiency. You may decide to change your altitude to reduce ground speed loss from winds at your cruise altitude. A smart budget flier checks wind conditions at various altitudes before and during flight through the FAA Notices to Airmen (NOTAMS) and Pilot Reports (PIREPS) on the communications radio. A 2000-foot change in altitude can often change a side-quartering wind to a tailwind that will improve both ground speed and fuel efficiency.

You can also save fuel by planning your descent in advance of reaching your destination. If it took you 20 minutes to ascend to your cruise altitude at 500 fpm and you plan to descend at the same speed, simply start descending to pattern altitude about 20 minutes before you expect to reach your destination and you'll conserve fuel.

If you're flying into a controlled airport where traffic controllers tell you when and where to land, planning your descent can help you give them an estimated time of arrival (ETA) at your destination airport and cut down the time you must circle before you can land. Calling your intent to land early may clear the way for a fast and fuel-efficient landing.

As a student you put down your flaps and, if so equipped, landing gear early in landing. Now that you are an experienced pilot, you can fudge a little and save extra fuel by waiting for the last safe moment to put drag on your plane with flaps and gears. Of course, it will take a lot of fuel savings to pay for a belly landing or a broken nose gear from a too-fast landing if you wait too long. Practice makes efficient.

Finally, you can save fuel—and your engine—by minimizing taxi time on the ground. Air-cooled aircraft engines require high revolutions per minute (rpm) to stay cool. Taxi speeds don't offer enough rpm to cool your engine, so minimize the time you operate your engine on the ground and you will be dollars ahead.

The concept of fuel efficiency is simple: save fuel where you can, but not at the expense of safety. Plan ahead and carry no more weight than you need. If you operate your plane by the owner's book, you'll both save fuel and enjoy flying on a budget.

Where to Put It

Parking your car is a simple matter. At work you just drive it into a paved area and place it between evenly spaced lines. At home you drive it into the garage and shut the door. No problem.

Parking your plane is not so easy. First, the plane must be parked at or very near an airport runway so that it can come in and go out on the wind. It must also be protected from the elements. That it can be with either a tie-down or a hangar.

A tie-down allows you to attach special cables from the wings and fuse-lage to anchored rings in the ground. It will only protect your plane from the wind. It offers no protection from sun, snow, or cold.

A hangar can protect your bird from all the elements and offer you a protected storage area reasonably safe from man and nature. The problem is that cost and availability may make a hangar difficult for the budget flier to justify.

Which type of storage you eventually choose for the plane you own, share, lease, or build depends on your plane and your flying budget. Older, fabric-covered planes can have their lives shortened by open storage such as with an airport tie-down. Of course, you must weigh that potential loss against the actual high cost of renting or buying a hangar.

In considering a tie-down at a local airfield, ask yourself questions such as these:

What is the cost of this airport's tie-downs?
Is there a lower rate for long-term tie-downs?
Is the plane tied down to concrete-anchored hooks or wooden stakes?
Are available tie-downs on grass, gravel, or paved areas? How close are they to taxiways?
How easy will it be to get my plane to a taxiway?
Is the tie-down visible from the flight office to reduce the chance of theft?
Is this an area of high winds that will require special anchors?
Is the weather such that I will need window, prop, engine, or wing covers to reduce the heat or cold? Will I have to preheat the engine in winter?
Has this airport had problems with vandals or theft in the past? What responsibility does the airport take with the rental of a tie-down?

For some budget fliers, tie-downs are the most economical and practical storage available; for others, they are an expensive headache. Review all of the factors before you decide.

Hangars are simply large metal storage buildings for housing airplanes. There are two types:

Common hangars
T hangars

A common hangar offers a sheltered area where many planes can park out of the elements. They are the least expensive because they are shared by many planes. The problem is that if the hangar houses six planes, yours is bound to be in the back and getting it out will necessitate moving at least half the planes.

A T hangar solves the access problem at greater cost. A T hangar has stalls shaped like T's so that planes—also shaped like T's—can be backed in from both sides of the hangar yet still be separated by common walls. T hangars cost more to rent than common hangars.

When selecting either hangar, look for one that is or can be heated in cold weather to reduce warm-up time. If nothing else, can it be heated by your car heater parked nearby and run through flexible ducting? Also check the proximity of food, weather information, and toilet facilities. Ask the FBO about security. Are the hangars patrolled regularly? Who supplies the locks? Does the office have a key? Who is responsible for loss or damage?

Some smart budget fliers will bring down the cost of storage by combining household or business storage with plane storage. That is, they rent T hangars and then use part of the enclosed space for storing furniture, household goods, business records and equipment, or other things needed only occasionally. Check with both your FBO and home insurance agent to make sure that goods stored will be covered.

Location of your storage will also dictate what you use. If no hangars are available at fields within 30 miles, you may be confined to tie-down storage of your wings. Contact the FBOs in your area in person or by phone to find out what's available, how much it will cost, and how much climate and financial protection the storage offers.

The amount of flying you do also will help you decide what type of storage you need. Daily flying in an area of tropical weather will suggest that tie-downs are the best way to go, whereas low usage during cold winter months may require you to rent a heated hangar even though the annual cost is high. So is the cost of airplanes.

As you shop for a home field, you should consider the supporting services offered. Is aviation fuel readily available and for sale during the hours you fly? Does the fuel have the right octane rating for your plane? Is low-cost maintenance available on the field? Is there more than one repair service to encourage competition and lower prices? Are parts for your plane available nearby? Can you find an instructor at the field who can help you when you want to upgrade your license and skills on a budget? Are the majority of planes and pilots at the field budget fliers like yourself, or does the field cater more to the business flier and twin-engine plane? Can you rent the plane you want at the field or do you need to go to a field where you are not so well known?

The smart aviation consumer who is willing to do a little digging and asking can save hundreds of dollars each year when purchasing aviation insurance, fuel, and storage by thinking like a budget flier.

8 KEEPING MAINTENANCE ON THE GROUND

If you have mechanical problems with your car, you may, if you wish, put it up on blocks and fix it yourself.

Not so with your airplane. For the protection of yourself, future owners, your passengers, and people on the ground, the FAA requires that all repairs on production airplanes be done under the supervision of certified mechanics.

That ruling not only makes pain for your pocketbook but also reduces your chance to understand your wings from the inside out. You don't want to fall out of the sky, but not even certified mechanics issue unlimited warranties on their work. Your best insurance is your understanding of the machine you fly and how it works.

How does the budget flier fit into all this? How can you reduce the cost of maintenance and repairs without reducing safety?

With knowledge. You must understand the rules and regulations governing aircraft maintenance and repair. You must know what you can legally do to keep your plane flying safely. You must also know how to do it. You should take the time to study parts and repair manuals for your craft. By doing so, you will lower the cost of going up while you increase your enjoyment of flying. Legally, you can:

Do your own basic maintenance suggested by Federal Aviation Regulations Part 43.

Make all of your own repairs under the supervision of a certified A&P mechanic.

Maintain and repair your plane if it falls in the experimental, amateur-built classification.

Depending on your mechanical abilities and your willingness to learn, you can cut your maintenance and repairs at least in half. You can handle your own required maintenance, preventive maintenance, and engine or airframe repairs in your spare time and save your cash for flying safely.

The FAR Side of Flying

To be up in the sky legally, an airplane must have an FAA Certificate of Airworthiness, which states that the plane is safe to fly when it is maintained according to Federal Aviation Regulations (FARs). A factory-built plane will have the certificate when it's born.

A home-built plane, when built under FAA inspection as outlined in Chapter 6, earns the designation *Experimental Aircraft.* That means the design has not been FAA-approved and is subject to certain restrictions. The restrictions depend on the craft and its design, but they are usually limited to noncommercial flying within a specific radius of the home field for the first 50 or more hours of flight. Approval of your craft earns an FDA Design Type Certificate (TC).

Before you pick up your first torque wrench to work on the TC'd airplane, you should read the instructions—the government's list of what you can and can't do to a production airplane—in FAR Part 43, Appendix A. It says that you can perform the following preventive maintenance measures on any plane you own or fly except one used as an air carrier:

Removal, installation, and repair of landing gear tires
Replacing elastic shock absorber cords on landing gear
Servicing landing gear shock struts by adding oil, air, or both
Servicing landing gear wheel bearings, such as cleaning and greasing
Replacing defective safety wiring or cotter keys
Lubrication not requiring disassembly other than removal of nonstructural items such as cover plates, cowlings, and fairings
Making simple fabric patches not requiring rib stitching or the removal of structural parts or control surfaces
Replenishing hydraulic fluid in the hydraulic reservoir
Refinishing decorative coating of fuselage, wings, tail group surfaces (excluding balanced control surfaces), fairings, cowling, landing gear, cabin, or cockpit interior when removal or disassembly of any primary structure or operating system is not required
Applying preservative or protective material to components where no

disassembly of any primary structure or operating system is involved
and where such coating is not prohibited or is not contrary to good
practices

Repairing upholstery and decorative furnishings of the cabin or cockpit
interior when the repairing does not require disassembly of any primary structure or operating system or interfere with an operating system or affect primary structure of the aircraft

Making small, simple repairs to fairings, nonstructural cover plates, cowlings, and small patches and reinforcements not changing the contour
so as to interfere with proper airflow

Replacing side windows where that work does not interfere with the
structure of any operating system such as controls and electrical
equipment

Replacing safety belts

Replacing seats or seats parts with replacement parts approved for the
aircraft, provided the work does not involve disassembly of any primary structure or operating system

Troubleshooting and repairing landing light wiring circuits

Replacing bulbs, reflectors, and lenses of position and landing lights

Replacing wheels or skis when no weight and balance computation is
involved

Replacing any cowling not requiring removal of the propeller or discon-
nection of flight controls

Replacing or cleaning spark plugs and setting spark plug gap clearance

Replacing any hose connection except hydraulic connections

Replacing prefabricated fuel lines

Cleaning fuel and oil strainers

Replacing batteries and checking fluid level and specific gravity

That's the list. Uncle Sam says you can take any of those preventive
maintenance steps without supervision or having to have the work certified.

There are many other maintenance steps and repairs you can handle *if
you work under the supervision of a certified mechanic who is qualified to
inspect and certify your work.* That's the catch 22 that allows the budget flier
to save hundreds of dollars in labor each year by doing his or her own
repairs—or at least part of them—and then having the finished work inspected and logged off in the aircraft logbook. If you know how to perform
engine and airframe maintenance and repairs or are willing to learn, you can
cut the cost of flying while you become better acquainted with your wings.

Required Maintenance and You

Beyond the preventive maintenance that you can do to keep major repair costs at a minimum, the FAA requires other maintenance on a regular basis. The most common examples are the *annual* and the *100-hour inspections.*

Once a year an airplane not used for commercial flights must be inspected for overall condition and assurance it is airworthy. This is the so-called annual. If your craft is a production plane with an Approved Type Certificate, the annual should be performed by a certified mechanic with an inspection authorization (IA) rating. If repairs are necessary, you can have the mechanic do them. Alternatively, you can do them yourself under the mechanic's supervision and final inspection. If your plane has an Experimental, Amateur-Built Airworthiness Certificate, you can legally perform and sign off your own annual inspection.

If your plane is used commercially, you must also subject your wings to a 100-hour inspection to ensure the safety of your passengers and cargo. The inspection covers the same ground as the annual but a certified A&P mechanic, rather than an IA-rated mechanic, can sign your aircraft log if you wish. Of course, the 100-hour inspection isn't required if you don't fly for hire.

Sometimes the FAA issues Airworthiness Directives (ADs), to owners of planes of a specific model, requiring inspection or repair of certain parts or equipment. This can also be classified as "required maintenance," but we'll cover it later in the chapter under modifications and repairs.

Help!

If you've decided to join the budget fliers who do some or all of their own repairs, there are many ways you can make your job easier.

First, you can begin building your maintenance library. Start with a copy of your plane's owner's manual. If your plane doesn't have such a manual, you can buy one from either the manufacturer or one of the larger aircraft parts houses such as Wag-Aero and Univair (see Chapter 6). You can usually buy other service manuals and valuable parts from the same sources. As an example, the total cost of a service manual, parts list, and owner's manual for the popular Piper J-3 Cub is less than $20.

Your library should also include books like *Aircraft Maintenance for the Private Pilot* by Kas Thomas, *The Aircraft Owner's Handbook* by Timothy R. V. Foster, *Lightplane Owner's Maintenance Guide* by Cliff Dossey and *Lightplane Construction and Repair* by Al Snyder and William A. Welch.

These books are available through most large book stores and aviation book sellers.

Next, find at least two reliable sources for parts for your craft. The following are the most common sources.

Factory-authorized dealers have parts you can't find anywhere else. Because they specialize in parts for specific brands of craft, they must often charge more than other sources. Often, though, the expert assistance they include can be worth the extra cost.

Aircraft parts houses handle the most popular parts for a wide variety of planes from old Swifts and Ercoupes to new Beech Bonanzas. You can find them by asking your FBO or your mechanic or by reading aviation magazines. Many houses offer valuable parts and supplies catalogs at little or no cost to potential customers.

Individuals can also be a good source of parts for your plane. You can find out what's for sale and what's wanted through aviation publications, airport bulletin boards, and flying clubs. If you're restoring a Luscombe or Navion or other out-of-production plane, there's sure to be some individual with the part you need.

Aircraft junk yards are also an excellent source of used and hard-to-find plane parts. Check the phone book and local repair shops in your area to see if there is a plane junk yard nearby that can serve as a source of parts for your plane.

Finally, the smart budget flier cultivates the friendship of a qualified A&P mechanic. Most mechanics will cooperate with pilots who can handle some or all of their own repairs but need certified mechanics to thoroughly inspect and sign off work done by the planes' owners. Some mechanics may even rent you space in which and tools with which to make your repairs at a lower rate than if they had to supply both space and labor. Talk to fliers and FBOs in your area to discover mechanics who will work with the budget flier.

Even if you are only slightly mechanically inclined, you can save money by working with an understanding mechanic. You can save the mechanic's valuable—and costly—time by preparing your plane for the annual inspection or overhaul. You can remove cowling and covers, fabric and other parts that give the mechanic better access to the area to be inspected or repaired.

In addition you can be on hand to help a consenting mechanic with tools, parts, and assistance on minimal jobs. By doing so, you will both save on labor charges and learn your way around your plane. You will soon find yourself confident enough of your knowledge to do more and more of your own repairs—on a budget.

Improving Your Performance

No two planes are alike. Nearly every plane ever made has acquired a modification somewhere along the line. It may have amounted to changing the prop, enlarging a window, adding wing-tip tanks, or increasing the engine's horsepower. Pilots, it seems, just can't leave things alone. They are constantly reaching for the ultimate plane.

Evolutionary modifications can often be made by or with the help of the pilot. Once you have decided what modifications you want, you can assist or work under the supervision of your friendly mechanic to make the changes and have them certified in your logbook. An amateur-built or home-built plane can often be modified without outside help.

There are four ways to modify or improve your plane. They include:

Performance modifications made to improve engine horsepower, cruising speed, rate of climb, or STOL capabilities.

Efficiency modifications to make the plane more practical for the owner. They range from converting a tricycle gear into a taildragger or seaplane, increasing the available load or seating, or increasing range with wing-tip tanks to upgrading the fabric covering.

Aesthetic modifications include changing the design of the wings, fuselage, or tail of your plane, beautifying the instrument panel, or adding a special paint job.

AD modifications include changes in your plane to comply with Airworthiness Directives issued by the FAA. They can include installing redesigned door latches, replacing specific parts with more airworthy parts, or changing parts of the fuselage.

In each case you can have the modification made or you can make it yourself, depending on the complexity of the job and your own flying budget. A STOL package can run between $1000 and $6000 or more depending on the plane being modified, the changes being made, and the company doing the work. Upgrading an engine and prop can cost $5000 to $10,000 on many planes, and adding wing-tip tanks carries a price tag of between $1000 and $3000.

Or you can do it yourself. In any case, the person or company making the modification must take care of the extra paperwork required to notify the

FAA that modifications in your plane have been made. That usually means filing for a Supplemental Type Certificate (STC) on FAA Form 337. The filing must include drawings and data on the changes you propose and the purpose of the changes. If you're considering doing the work yourself, talk with the local FAA GADO office about how to do it. Also check with local mechanics to get an estimate of the cost of having someone else do it.

Saving on Repairs

The airplane is actually a simple machine. Most planes have fewer moving parts than the car you drive; they usually operate on fewer cylinders; they lack a transmission or differential; and they are highly fuel-efficient.

They are also very critical. If you have engine trouble, you can't simply pull over to a cloud, stop, and pop the hood. Maintenance and repairs must be done on the ground. It's that critical factor that makes both the FAA and smart pilots sensitive about keeping planes in top shape and making repairs before they are necessary.

Yet many pilots do much of their own repair work to both save money and learn more about their flying machine. You can, too, if you're willing to build your own plane under the experimental, amateur-built classification or work under the supervision of an IA-rated mechanic.

Aircraft engines are highly efficient; they have a normal time between overhauls (TBO) of 1200 to 2000 hours. You can extend the TBO of your engine with preventive maintenance and by making sure that your engine receives proper lubrication, good gas, and a clean spark.

You can also do part or all of your own engine overhaul. Some budget fliers cut costs by removing the engine from the plane for the mechanic to lower labor time, and others actually assist in the rebuilding process.

There are two types of engine overhaul:

Major overhaul
Top overhaul

The TBO for your plane is an *estimated* time between *major* overhauls. A major overhaul consists of a complete disassembly of all parts, a thorough inspection for cracks and wear, replacement of parts not within tolerances, reassembly, and a running test of operation.

The top overhaul, which is done between overhauls at intervals recommended by the manufacturer or your mechanic, includes inspection and replacement of parts outside the crankcase—valves and cylinders primarily.

An ad for a plane you're considering might read "100 hrs. STOH," which means it's been 100 hours of flying since *top overhaul* as opposed to SMOH (*since major overhaul*).

You must also consider airframe repairs to save on your flying budget. They are usually easier than engine repairs for most pilots, and they offer an opportunity for real savings.

The most common airframe repair done by budget fliers is re-covering a fabric plane. Because of weight considerations, most planes more than 25 years old—and many planes not that old—were covered with fabric stretched and glued over the airframe. The older planes were covered with grade A cotton, which has a five- to seven-year life expectancy. Today the newer planes are covered with, and most older planes are being re-covered with, a material called Ceconite, a woven Dacron polyester fabric that lasts about 10 years.

Re-covering must be done either by or under the direction of a certified airframe mechanic. A Ceconite re-covering envelope will cost between $200 and $500 depending on the amount of surface. Dope and labor are extra. If you're planning to supply the labor, you can read up on the re-covering process in Joe Christy's *How to Install and Finish Synthetic Aircraft Fabrics* and *Modern Aircraft Re-covering* written and published by Airtex Products, a firm that specializes in aircraft interiors and exteriors (see Chapter 6).

If your craft is metal or wood, you can update its skin with a new paint job. The job is labor-intensive, so the budget flier can save many hundreds of dollars by doing the work instead of hiring someone else to do it. It's a matter of stripping all of the old paint off, cleaning the surface, and applying a corrosion-resistant primer followed by a urethane primer, the main color, and then any secondary colors.

Repairs to aviation electronics are something else again. In many well-equipped older planes the value of the avionics can come close to that of the plane itself. Yet the avionics are much more complex. Repairing navigation and communications equipment is knowledge-intensive rather than labor-intensive, and unless you have a solid working knowledge of solid-state electronics—and an FCC second-class radiotelephone license—it's best to leave repairs of the little black boxes to the experts.

However, you can lower the cost of needed repairs by shopping for value and price as you would for any other piece of the aviation puzzle. You can consider the purchase of each new unit closely to decide whether it will help you with the safety and pleasure of flying enough to justify the cost. You can compare the costs of buying new, used, or rebuilt equipment and repairing your own equipment. You can ask among friends for qualified avionics technicians who would be willing to do the work in their spare time for a reduced labor rate or in exchange for a skill that you have earned.

As a budget flier, you will keep things simple and shy away from the more sophisticated avionics designed as rich men's toys: flight directors, autopilot, aircraft weather radar, and airborne telephones. You will want to fly the plane with your own hands rather than through the mind of a sense-deadening electronic gadget. Flying is a sensation.

Choosing Your Tools

If you've decided to do some or all of your own maintenance and repairs, you'll need a basic set of tools for both the mechanics and the airframe. If you're handy, you may have many of the tools already on hand. Otherwise, you can purchase the tools, rent them, or borrow them from a certified mechanic or another budget flier.

Here's a list of basic mechanic's tools you will need for many aircraft repairs:

$1/4$-in. square-drive socket set
$3/8$-in. square-drive socket set
$1/2$-in. square-drive socket set
Combination wrench set ($5/16$ through 1 in.)
Standard screwdriver set
Phillips screwdriver set
Punch set (centerpunch, driftpin, and cold chisel)
Hacksaw and blades
Pliers (slip-joint, diagonal cut, long nose)
Torque wrench

Other tools you may need include:

Aircraft jacks
Magneto timing light
Engine drain oil can
Cable tensionmeter
Nibbling tool (for cutting sheet metal by hand)
Sheet metal brake (for bending up to 16-gauge sheet metal)
Aircraft tube bender
Riveter and riveting set
Rivet cutter

Aircraft welding outfit
Airbrush painting kit
High-speed air sander
Aviation snips
Instrument-hole-cutting tool
Volt-ohmmeter
Soldering gun

You can economize by renting most of these special tools instead of purchasing them unless you expect to make a great many repairs. With them you can handle nearly any basic repair to your plane and save money.

You can increase both your safety and your enjoyment of the skies by legally handling many of the repairs necessary for the active aircraft. You can better understand yourself and your wings as you discover maintenance on a budget.

9 CUTTING CLIMBING COSTS

You're hooked!

You've learned how to fly on a budget, and you've begun your discovery of the sky. You've overviewed your world from 3000 feet, and your perception of places and things within it has changed.

You're ready to take whatever the next step is to expanded awareness and pure flying fun. You've decided to improve both your wings and your skills to enjoy the sensation of flight more deeply.

You're also aware that refinements of flight will cost you more money. You realize that you are going to have to improve your aviation consumerism even more in order to grasp more of the pleasure of flying on a budget.

But you're willing to learn. You've been hypnotized by flight, and you're ready to widen your horizons as other successful budget fliers have done.

Getting More for Less

As you grow in aviation, you will discover new ways to get more for your flying dollar. You'll find new techniques for cutting the cost of going up.

The first step to improving your business or pleasure flying within a budget is to review your budget for weak and strong points. You must analyze your budget to make sure that each dollar is spent wisely and to decide where wasted dollars are going—and where they can be channeled.

Next, you should decide how you want to improve your flying skills. You may want to upgrade your pilot's license, learn aerobatics, study mechanics or aerodynamics, or sharpen your skills in competition.

Then you can study the best way to upgrade your wings. You'll look at choosing your next plane by weighing your needs and budget. You'll rethink the question whether to buy, rent, or build your next plane. You'll learn how best to sell your current plane if you need to do so.

You'll also consider how to upgrade the wings you have to meet your increased needs and skills. You'll study the benefits and costs of STOL; you'll consider more sophisticated avionics black boxes and how to buy them. You'll look at the advantages and disadvantages of hanging onto the wings you have.

In short, you'll discover and adapt dozens of ways for getting more flying for less money.

Dusting Off Old Records

You've been a budget flier for awhile now. Whether you've purchased, shared, rented, leased, built, or rebuilt your first set of wings, you've probably kept records to stay aware of the cost of flying. It's time now to look back over them as you would at the closing of an old year. If all you have is a pile of receipts and a special checking account for your flying, now's a good time to put your records in permanent form with a flying budget book.

First your budget book will have facts and dates on your flying. When did you start? How long did it take to earn your license? How much did it cost? How many hours have you flown since then?

Then it will break your flying time down further. How many total hours have you flown? In what time period? What is your average flying time per week? Per month? Each year? What type of equipment have you flown?

What uses have you made of flying? What percent of your total time in the sky was for business? For pleasure? Of your pleasure flying, how much was personal and how much was family? Did you use your plane for a vacation?

Looking at costs, what have your fixed costs been? If you purchased your plane alone or in partnership, what was your down payment? What are your monthly payments? How much does insurance cost? Storage? Maintenance? If you belong to a flying club, what are your monthly dues? If you built or rebuilt your wings, what was the cost of your project?

What were your variable costs during your trial period? What was your total fuel bill? How many miles per gallon did you get with your wings? How much oil was used? What were the costs? Were they higher at some airports than at others? Did you correctly estimate your maintenance reserve? What maintenance was necessary? What was the cost? How much of the maintenance were you able to do ? What was the cost of parts?

Finally, it's time to total it all up and decide how efficient you've been as

STOL (short takeoff and landing) aircraft are increasingly popular. This Robertson STOL conversion of a Cessna 182 Skylane RG (retractable gear) can increase the plane's cuise speed by 3 miles per hour and cut the normal takeoff distance over a 50-foot obstacle from 1350 feet to just 815 feet. It can also lower final approach speed from 70 to 47 miles per hour. *(Robertson Aircraft Corporation)*

a budget flier and, just as important, how you can improve your record. You can start by comparing the bottom lines: how much did flying cost me and what did I get for it? As an example, if you operated your plane 37 hours in the last six months, you averaged just over 1.4 hours per week. If your costs for the six-month period totaled $870, your per-hour cost of flying was about $23.50. How does that compare with renting or leasing the same plane in your area? Were you able to take advantage of tax savings through business use? Can you reduce the costs even more?

Take a look at your initial budget again. How much did you project that flying would cost when you first got into it? How good was your guess? What can you do to improve your flying budget?

Here's how smart budget fliers trim the fat off their flying budgets to get more time in the sky:

Reconsider ownership   If you own your wings by yourself or with a part-partner, review the costs and benefits to make sure that you are getting the most for your money. Compare ownership with other methods of getting up in the sky: rental, leasing, flying clubs. Or if you don't own, consider that method.

Review costs of flying   Now that you know what your flying costs are, look over what you've spent on flying to see what can be trimmed. Are

You can upgrade your wings into a float plane with conversion kits available from many manufacturers and increase the utility of your airplane. (*Cessna Aircraft Company*)

membership dues in a flying club worth the cost when compared with rental fees through an FBO? Is your plane fuel-efficient? Does it demand high-cost maintenance or insurance? What can you do about it?

Modify the flying budget   You may now decide that your initial flying budget is too high or too low for your current interest in flying. How much do you now feel flying is worth to you—more or less than your initial budget? What type of flying held the most interest for you? Cross-country? Flying for business? Sports flying? How can you pack more of that type of flying into your budget?

Reviewing your flying budget and how you spend it can help you redirect your resources to the style of flying you enjoy most.

Courting Your Second Bird

As you grow as a budget flier, your needs and desires will grow. Soon the simple bird with which you first took to the sky will seem inadequate. Your Aeronca Champ may seem too slow to get you to your desert hideaway on Friday afternoons. Your Piper Super Cub may not have enough seating to take the whole family along. Your Cessna 150 may not have the instruments you need to use your new IFR rating and fly year-round.

Then it will be time to begin shopping for your next plane. How? Much in the way you chose your first bird. You will take a look at your current needs, your budget, and your expected utilization, you will compare cost and value, estimate fixed and variable operating costs, flip a coin, and decide.

Revising your flying needs means reviewing the basic differences between planes and deciding which features fit you best. As listed in Chapter 3, the points to consider include seating, load, power, fuel economy, cruise speed, range, configuration, landing gear, propeller, and avionics. As an experienced budget flier you might decide a four-place plane is more efficient than your two-seater. Or you might find an older plane with retractable landing gear for greater fuel efficiency. In any case, comparison shopping is as important in selecting your second plane as it is in your first.

To help you with your decision on which wings to buy, rent, or rebuild, Chapter 12 offers a dozen of the Best New Aviation Buys complete with photos and performance and economy figures. Chapter 13 covers Best Used Aviation Buys and introduces you to a dozen of the planes most popular with budget fliers.

Your budget will help you decide *how* you can best gain your wings: buy, rent, lease, build, or rebuild. By comparing the cost of your flying with the cost of other models of flying, you can discover the most efficient and least costly way to go flying. After owning a plane in a partnership with three other budget fliers, you may decide to build your own wings from a kit to gain both the experience and the lower cost of ownership and maintenance.

Alternatively, you may decide to purchase or lease a newer plane in order to utilize it in your business and thus be able to deduct much of its operating cost as an expense. Now that you have experience with flying on a budget, you are more aware of your own needs and financial abilities. You can easily compare them with the aviation market to find the best plane and the best way to use it.

Shedding Your Wings

Once you've made the decision to upgrade your wings, you have another problem: how to get rid of your old ones. There's no problem if you're renting or you belong to a flying club, because you don't own anything. If you're a member of a partnership, you can talk with the other partners to work out arrangements. In most cases your partners will want to have final approval on anyone who buys your share and may even want first opportunity to purchase it themselves.

If you've built your own plane under the experimental, amateur-built classification, you should be aware that the next owner will not be able to

Automatic direction finder (ADF) receiver. *(General Aviation Electronics)*

do the same kind of maintenance on the bird that you did. The second owner will not have built more than 51 percent of the plane and so will have to get a new Certificate of Airworthiness from the FAA. If the certificate is issued, the new owner will have to have maintenance and repairs done by a certified mechanic just as though a production plane were involved.

If you now own a plane and you're ready to sell it to purchase your new wings, there are a number of ways you can go:

Sell it yourself   You can place an ad in one of the local newspapers or national flying publications. This usually takes more time than other methods, but it can net you a better price.

Sell it through a broker   An aircraft broker operating out of one of most of the larger general aviation airports can accept your wings to sell consignment. That is, the broker will show your plane with other planes. If the plane is sold, you will pay the broker a commission, usually about five percent. The length of time to sell your craft is often the same as if you did the selling yourself, but the broker handles the paperwork and transfer of title.

Trade it in   The dealer or individual you are purchasing your second plane from may buy your plane from you, but usually at a lower price than you could get if you sold it yourself or through a broker. The difference is time. By trading your old plane in, you are streamlining your purchase. You will not be stuck with either two planes or no plane at all. You exchange value for this convenience because you usually get little more than the wholesale price for your wings in a trade.

Sell it to a reseller   Flying publications advertise "Cash for your plane!"

ADF bearing indicator. *(General Aviation Electronics)*

A reseller will purchase your plane at wholesale value and take the risk of reselling it at market value. You get quick cash for your plane, but it's less cash than you would earn by selling the plane yourself or selling it through a broker.

Once you've decided to sell your wings, start preparing for the sale. Get your plane in top shape; clean it out; wax it; update the logs; check on the balance of your aircraft loan, if any; and talk with your banker or dealer about how to transfer title once the sale is made. You'll find most of the information you need on the transfer in Chapter 4, on buying your wings.

You can set your price by using both aviation price guides mentioned in Chapter 4 [*Aircraft Price Digest, Used Aircraft Price Guide,* and the *ADSA* (Aircraft Service Dealers Association) *Blue Book*] as well as by shopping for the plane you're selling. That is, see what others are asking for comparable planes. Talk with aircraft brokers, private parties, and FBOs to see what's for sale, how fast it sells, and what price sells it quickly.

Part of selling is financing. If you can help your buyer obtain financing for the purchase, you can move the sale faster and often get a slightly better price for your wings. Talk with the agency or broker who is financing your wings to see if there's a willingness to have a qualified purchaser assume or refinance the loan on your plane. If you own your plane outright, consider financing the purchase yourself. The down payment can be used as part or all of your own down payment on your next plane, and the interest you earn will help defray the interest you'll have to pay whomever finances your second set of wings. Owner financing can often make the difference in selling a plane, and it can command a higher price than having the buyer use a bank or aircraft lender.

Remodeling Your Wings

Then again, you may decide that your current craft is the greatest thing in aviation since the Wright Flier. It may have just the right performance and economic characteristics for the type of flying you do. It may be efficient and reliable and require little maintenance. It may be paid for.

Rather than get rid of a good thing, you may decide to simply make it better. There are many things you can do to improve your wings without trading them off for an unknown bird:

Change a tricycle gear into a taildragger gear for better ground clearance on unimproved runways

Change a taildragger gear into a tricycle gear for greater visibility on the ground

Install larger fuel tanks or wing tanks to extend flying range

Install a STOL kit to improve short-field operation

Replace cotton fabric cover with Coconite, Razorback, or other longer-lasting covering

Add extra instruments or redesign the instrument panel

Install better communications or navigation system

Replace cruise propeller with climb or other specialized prop

Install floats and make seaplane conversion

Install skis for landing on snow

The most popular improvements made by pilots seem to be in avionics a field that is becoming highly sophisticated and inventive. The general aviation pilot now has available black boxes that the airline pilot of 20 years ago would like to have had aboard. Although the budget flier must be selective in what he or she chooses to buy, even an economy-minded aviator can take advantage of the gadgetry that can make flying both safer and more enjoyable.

Area navigation systems (RNAV) combine with your DME or DCE to create a VOR station wherever you want one and let you fly right to it. You may find a used plane with an RNAV included that will let you enjoy advanced avionics at a low price. *(Collins Division of Rockwell International)*

Here are some of the black boxes the aviation consumer can choose from today:

DME   Distance-measuring equipment simply tells you how far you are from the VOR (VHF omnirange) navigation station you're monitoring on your NAV/COM discussed in Chapter 3.

RNAV   The area navigation system is a fascinating invention that takes information from your VOR and DME and seemingly moves the VOR station you're monitoring anywhere you desire. Rather than use a VOR 20 degrees off your destination, you can use your RNAV to move its signal to your destination and then simply fly right to it.

ADF   An automatic direction finder operates like the VOR navigation system in your NAV/COM radio except that it uses standard broadcast stations (550 to 1650 kilohertz) rather than VOR stations as direction finders. The ADF is handy as a backup to your VOR system, especially in less populated areas where VORs are widely spaced. VOR signals don't travel as far as lower-frequency standard AM radio station signals do.

RMI The radio magnetic indicator is simply an ADF combined with a compass to show you the way to the AM station you're monitoring for navigation.

None of those black boxes are necessary to budget flying. The only instruments required for the average plane are the airspeed indicator, altimeter, magnetic direction indicator, tachometer, oil-pressure gauge, oil-temperature gauge, manifold pressure gauge, and fuel indicator. Unless you get into controlled airspace around large airports, you really don't have to have communications and navigation equipment—legally. However, flying for fun or business can be made more enjoyable and a lot safer by the addition of at least a few optional instruments. As a smart aviation consumer, the choice is yours.

If you do decide to add to the basic instruments, here's how you can get your money's worth. First, take a look at the type of flying you do. Is most of your flying within your county, or is much of it cross-country business flying on a deadline? Pleasure flying doesn't require sophisticated navigation systems, but business flying can.

Second, take a look at what's available. Review the basic instruments to decide which will best fit your needs. Is a VOR sufficient, or should you consider an RNAV system? Is an ADF backup necessary? (It is if you are going to do any serious IFR work.) Talk with other pilots, especially those who do the same type of flying as you do.

Third, compare cost and value. What will the equipment cost you? How much will it be worth to you? How often will you use it? Will its purchase fit comfortably into your budget?

Finally, start shopping. Compare brands and models, features and benefits, new and used. Prices vary greatly. As an example, a new RNAV system starts at about $2500 and climbs to over $4000, whereas a used or rebuilt RNAV may be purchased for $1200 by the smart aviation consumer.

A Helping Hand

There's one final way in which you can almost be sure you get your money's worth in aviation: keep informed. That is, stay current on what's happening in the world of general aviation and especially in aviation consumerism. An informed buyer is a smart consumer.

Toward that goal, the budget flier should subscribe to one or more of the many general aviation publications available, join a pilot's association, and talk with other pilots about problems and praises. The leading aviation magazines include:

Air Progress, published monthly for private aviators, offers articles on planes, techniques, careers, and equipment for both the business and pleasure pilot. 7950 Deering Avenue, Canoga Park, CA 91304.

The AOPA Pilot is published monthly by the Aircraft Owners and Pilots Association for its members. 7315 Wisconsin Avenue, Washington, DC 20014.

Flying is published monthly for private and commercial pilots involved with general-aviation aircraft for business or pleasure. 1 Park Avenue, New York, NY 10016.

General Aviation News is a biweekly tabloid newspaper on general aviation. Box 1094, Snyder, TX 79549.

Plane & Pilot is a monthly magazine on general aviation that is of special interest to the new and budget flier. 606 Wilshire, Suite 100, Santa Monica, CA 90401.

Private Pilot is published monthly for owner-pilots of private aircraft, student pilots, and others who aspire to attain additional ratings and experience. Box 4030, San Clemente, CA 92672.

Over a quarter-million plane owners and pilots are members of general aviation's largest fraternity, the Aircraft Owners and Pilots Association (AOPA). The purpose of the organization is to represent general aviation to our legislators and the public as well as serve members with useful information and benefits. Membership dues include a subscription to *The AOPA Pilot* magazine, which has reports on regulation changes and new equipment developments, travel reports, and other information, a minimal flying-only personal accident insurance policy, and an informative newsletter. The AOPA also offers handbooks for pilots, training and ground school seminars, travel information, a chart service, and legal assistance at a nominal charge for members.

For further information on membership you can contact the AOPA by writing:

Membership
Aircraft Owners and Pilots Association
7315 Wisconsin Avenue
Washington, DC 20014

One of the most useful publications for the budget flier is *The Aviation Consumer,* which is published twice a month. It contains no advertisements; instead, it offers articles on products and services available to the pilot.

Some of those articles include: "The Aircraft as an Investment," "Weekend Cram Schools," "Avionics Survey Wrapup," and "Dealing with the Aircraft Mechanic." Subscriptions are relatively high because of the lack of advertising, but one article can often save you the price of the subscription. The subscription address is P.O. Box 972, Farmingdale, NY 11737.

The budget flier stays informed on how and where to get top value for each aviation dollar and cut climbing costs.

10 EARNING YOUR ADVANCED WINGS

Someday soon, flying your single-engine plane only on days when visual flight rules (VFR) apply is not going to be enough. You're going to want to break through the clouds and explore the sky above with your instrument rating. Or maybe you'll want to learn to fly a twin-engine craft, helicopter, glider, or balloon.

Maybe that time is now.

You can. Your private pilot certificate is the starting point for a lifetime of flying. With added knowledge, proficiency, and experience you can earn your advanced wings and fly in most types of weather, or fly other types of aircraft.

Best of all, you can do it all on a budget. It isn't going to be cheap, but if you're really interested in upgrading your skills and your licensing, you can apply the techniques of cost-conscious flying and earn your instrument rating or advanced certificate on a budget.

Bird Watching

On the ground and in the air, let's all be glad that the flier with a new private pilot certificate can't step into the cockpit of a Boeing 747 and fly it legally. It's too complex a bird to be handled by a pilot who lacks specific training and thousands of hours of experience.

Let's also be glad that the FAA licensing system offers the stepping-stones for the pilot who does want to eventually fly the 747—or we wouldn't have any airline pilots. The elite must work their way through a maze of certificates and ratings until they achieve the skills and license that best suit their flying goals.

Here's what the ladder of licensing looks like:

Certificates:
Student pilot
Private pilot
Commercial pilot
Airline transport pilot
Flight instructor
Ratings:
Instrument
Multiengine
Seaplane
Rotorcraft
Glider
Balloon

It works like this: first you earn your certificate; then you add ratings. With experience and proficiency you could someday hold a commercial license (actually a certificate) with an instrument rating. Or you could earn a flight instructor license with a rating that would permit you to teach single-engine and multiengine flying. The choice is yours.

The greatest majority of pilots—over 60 percent of AOPA members—go no further than their private pilot certificate. But nearly 75 percent of the members have gone beyond "single-engine land" and added advanced ratings to their certificate.

Why? Primarily to get more value from their flying. Increased skills and ratings mean greater utilization of their plane. Some go for ratings in order to get better jobs, but the greatest number of fliers elect to tack on an instrument rating to their certificate so that they're not grounded when instrument flight rules apply—during "IFR weather." They know that the costs of instrument instruction and an IFR-rated plane are higher, but as budget fliers they can see the value of greater utilization of time and equipment.

The instrument rating is probably the first one you'll be interested in earning. The FAA says that, to be eligible, you must hold a current private or commercial pilot certificate with an appropriate aircraft rating. That is, you can't go for an instrument rating in helicopters if you don't also hold a rotorcraft rating. You must also pass a written test on IFR and show a working knowledge of IFR traffic procedures, navigation, flight planning, and safety.

The FAA also wants you to have some experience before you begin viewing the clouds from above. A minimum of 200 hours of actual flying time, including at least 100 hours as pilot in command, 50 of which as cross-country flying hours, is required. The FAA also wants you to log 40 hours of instrument time and 15 hours of actual instrument flight instruction with an authorized instructor. These times can run concurrently, and you can probably count all the hours you logged while you were working toward your private pilot certificate.

Your flight test will include maneuvers flown solely by reference to instruments, IFR navigation, instrument approaches, cross-country flying during actual or simulated IFR conditions, and recovery from simulated emergencies.

Most pilots admit that going for an instrument rating isn't easy but that the effort is well worth the increased utilization. They can fly nearly year round and often fly above storms that ground the VFR pilot. Just as important, their confidence in themselves and in their craft increases with the addition of IFR skills and the knowledge that they can reinforce safety in the skies.

They are *smart* budget fliers.

A multiengine rating can be valuable, especially to the pilot working up the licensing ladder toward an airline career. Since most pleasure fliers stick to the simpler single-engine planes, the multiengine rating is used more by the business and professional flier.

Multiengine is an aircraft class rating. That is, it's a rating that qualifies a pilot to fly a specific class of aircraft: multiengine. This is the easiest rating to get. First, you must receive flight instruction from an authorized flight instructor who is rated for the rating you're going for. The FAA doesn't specifically say how many hours of instruction you need, but anything over 10 is usually acceptable. You'll also have to pass the flight test and an oral quiz that display your knowledge and skills in operating a multiengine craft safely and legally.

The seaplane rating is another aircraft class rating that can be valuable to many types of private pilots. If you plan to fly in an area where there are numerous lakes or large bodies of water, a seaplane rating can add hundreds of miles of runways to your world. For most types of planes, float conversions are available that add seaplane capability to a conventional plane.

Seaplane rating requirements are basically the same as the multiengine rating requirements. You must show the FAA examiner that you've logged seaplane instruction, answer a few questions, and pass a flight test.

You can also step beyond the airplane and go for *aircraft category ratings:* rotorcraft (helicopter), glider, and lighter-than-air (balloon). Each offers a new world of experience for the budget flier.

To earn your helicopter rating you must have at least 40 hours of flight instruction and solo flight time, including a minimum of 15 hours of solo time in helicopters. The 40 hours were earned when you picked up your private pilot license, so all you'll need is the 15 hours of instruction. The problem is that helicopter time is expensive. We'll talk about costs later.

For your glider rating you need to log 70 solo glider flights including 20 flights with 360-degree turns. Or you can log seven hours of glider solo flights including 35 ground-launched or 20 aero-launched flights. Or, if you already have a private pilot certificate, you can get your glider rating with 10 solo glider flights that include 360-degree turns. Then you'll be legal in the world of gliders and sailplanes, where the sky is silent and thermals are your best friends.

For a balloon (lighter-than-air) rating you'll need to log 10 hours in gas or hot-air balloons with at least six flights supervised by someone who has a commercial pilot certificate and free balloon rating. The flight requirements differ depending on whether you're using a gas balloon or a hot-air balloon with an airborne heater. As with all ratings and certificates, the specific requirements can be found in Part 61 of the Federal Aviation Regulations (FAR).

Time for a Commercial

The next step up from a private pilot certificate is the commercial pilot certificate. In spite of the name of the license, it doesn't authorize you to fly commercial aircraft like 727s and DC-10s. Rather, it's for the middle-ground pilot who flies passengers or property for hire in smaller aircraft *or* the pilot who just wants to be more proficient. To earn your commercial license you ·must learn advanced maneuvers—720-degree power turns, gliding spirals, and simple aerobatic maneuvers—that will make you a better pilot. About 30 percent of AOPA members hold a commercial license.

You'll have to go back to the books to pick up your commercial license. To pass the difficult written test you'll need to learn the FAR rules governing commercial pilots and more about basic airplane dynamics and airplane operation theory. The last category covers the use of flaps, retractable landing gears, controllable propellers, high-altitude operation, loading and balance, and airplane performance speeds. There will be additional topics and questions if you also want to go for a glider, rotorcraft, or lighter-than-air rating.

You'll also have to display flight proficiency with advanced flying techniques including precision approaches, slow flight, and the use of retractable landing gears and controllable props.

Finally, you need some hours on your log—250 or more. They must include 100 hours in powered aircraft, 100 as pilot in command, and 50 hours of flight instruction (including 10 hours specifically for instruments and 10 for the commercial certificate). Again, your private pilot training hours can be counted toward fulfilling these requirements.

It is almost mandatory that you get your instrument rating before at least at the same time as you get your commercial license. Without the instrument rating your commercial license will limit you to flying within 50 nautical miles of the airport and to daylight flying only. No fun.

With your commercial pilot certificate you can act as the pilot in command of an aircraft carrying persons or property for compensation—but only those of craft for which you hold the ratings. A rating for small aircraft (less than 12,500 pounds of maximum certificated takeoff weight) automatically comes with your certificate. Beyond that you'd need a rating for large aircraft, small turbojet-powered airplanes, and small helicopters, as well as other ratings for specific aircraft (L1011 TriStar, DC-3, 747).

As the Pro Flies

The top step on the licensing ladder is the Airline Transport Pilot (ATP) Certificate, and the requirements are stringent. The applicant must be at least 23 years old and must pass a stringent medical examination and have an excellent medical history.

The ATP candidate will also have to spend more time at the books studying for the demanding and technical written test. Subjects include the fundamentals of air navigation and equipment, elementary meteorology with emphasis on weather forecasting and reading weather maps, radio communication procedures, and loading and weight distribution principles.

The candidate must hold a commercial pilot certificate and must have logged at least 1500 hours of flying time as a pilot, including 250 as pilot in command and other specific requirements. That's nearly nine months of 40-hour workweeks in the air. Obviously, the ATP is not for the pleasure or weekend pilot. It's a career goal.

It's not surprising, then, that less than 5 percent of AOPA members hold an ATP Certificate. It takes years of training and experience for career-minded pilots to reach this rung of the licensing ladder. The reward is the opportunity to work as an airline pilot once the right ratings are earned.

Airborne Teachers

Many budget fliers mix business and pleasure to keep the cost of flying down. They become teachers both to help pay their way through the air and to log air time. Any pilot can teach someone else to fly, but in order for a student to get a license, the license must be signed by an FAA-certificated flight instructor rated for that type of plane or craft.

To earn your flight instructor certificate you must first hold a commercial or ATP license and pass written, oral, and flight tests. You'll need to take courses in the learning process, elements of effective teaching, student evaluation, quizzing and testing, course development, lesson planning, and classroom instruction techniques, and you must have taken required ground schools and know the FAR.

Many pilots use their instructor certificate to give lessons on days off from their regular jobs and then use the extra money they have earned to fly their own machines. Others build hours toward the ATP license and a better job. Still others just want to make a fair living by helping others discover the thrill of flight as they have. In any case, the flight instructor certificate is a worthy goal for the budget flier.

Low-Cost Advanced Instruction

The principles of budget flying aren't limited to the pilot who spends Sunday afternoon wearing down the traffic pattern in an Airknocker. They can be easily modified to reduce the cost of earning and enjoying the advanced certificates and ratings.

Since instruction is the first and sometimes the costliest part of flying, let's take a look at a few ways budget fliers get their money's worth out of upgrading their skills.

Most budget fliers go for their instrument rating first and then, if they want to, go on to the commercial certificate. The most expensive part of earning your instrument rating is not the instruction, it's accumulating the required 200 hours of flying time. You'll be able to count most of the hours you logged while earning your private pilot certificate, but you'll still need 150 or more hours in the air. How?

The next chapter, Chapter 11, will have many answers on how to make your flying pay for itself through business flying, multipurpose flying, lease-back, profit ventures, and other methods.

In most cases, instrument instruction will cost no more per hour than your initial certificate instruction did. The only difference may be the higher rental charge for an IFR plane, but chances are the plane you took your training in was rated for IFR instruction.

As before, the best way to cut instruction costs to a safe minimum is to shop around. Decide how many hours of instruction and ground school you'll need and start calling around to FBOs and schools in your area for comparative bids. Of course, don't shop price alone. Ask about the equipment used, its age and condition, the experience of the instructor, the training aids available, the scheduling flexibility, the terms, and all the other factors that make up your instruction package. Then go out and visit the FBOs that sound the best. Visit their school and talk with instructors and students.

You can also save on instrument instruction, or other instruction, by using the 10 techniques offered to the beginning student in Chapter 2: be prepared, be receptive, relax, be retentive, ask questions, keep a journal, do your homework, use your imagination, hang around, and schedule lessons as closely as possible.

For the price of an hour of two of flying time you can take advantage of thousands of hours of flying experience through books. The most popular are Richard L. Taylor's *Instrument Flying,* Robert N. Buck's *Weather Flying,* Richard Collins's *Flying IFR, Flying Magazine's Guide for Instrument Flying,* and the FAA's *Instrument Flying Handbook.* They can help you understand both the concepts and techniques of flying IFR.

You can also find study aids for earning your commercial and ATP certificate, multiengine, seaplane, rotorcraft, glider, and lighter-than-air ratings at larger book stores, in aviation magazine ads, and through aviation book clubs.

The Price of Advancement

The cost of instruction can be reduced by as much as 30 percent by shopping around in your area and comparing prices, equipment, and quality of instruction. As an example, in one metropolitan area the cost of instruction leading to multiengine rating varied according to the FBO—$499 through one FBO, $395 from another, $389 from one, and $449 from yet another. However, the student with the higher certificate in mind must shop more than price to get the best buy.

Generally, though, you can expect your instrument rating to run about $2000. Your seaplane rating will be about $300, as will the glider rating, after your private pilot license is earned, and helicopter hours start at $40 and go up beyond $100.

The budget flier can often beat these costs with comparative shopping, preparation, and a good mental attitude.

There are many excellent schools that can offer the instruction you want at a low cost. Unfortunately, there are also other flight schools that teach you little more than how to pass the required test. Here, again, is where the smart aviation consumer must investigate before investing time and money in poor instruction.

Flying schools range from single- plane bush pilot schools to resident universities for aeronautics such as Embry-Riddle. Size is not necessarily the best criterion for choosing a school, and many budget fliers have found professional instructors and lower-cost programs at dirt fields far from the major cities. Comparison-shop. Read the ads. Send for literature. Visit the facilities. Make your decision to upgrade your flying skills with the consumer principles you've learned as a budget flier.

There will be more on how and where to get advanced instruction in Chapter 11, on making your flying pay for itself. Your local phone book and a talk with FBOs and other pilots will yield other sources of low-cost advanced instruction.

11 MAKING YOUR FLYING PAY FOR ITSELF

Flying still isn't cheap.

You can purchase a Spartan plane and equip it with only the necessities. You can make many of your own repairs. You can increase fuel efficiency. You can even upgrade your skills and your equipment economically. But flying is still going to take some of your hard-earned money.

The idea behind budget flying, though, is to keep the cost to a minimum while enjoying safe flying. Toward that goal, there are still many things you can do.

You can combine business and pleasure. You can take advantage of your wings by using them more in your profession. You can use your plane for vacations and as a springboard to new and exciting recreational activities. You can get greater use out of your plane.

You can also reduce the actual cost of flying by using your wings as a source of full- or part-time income. You can explore the many opportunities available to the private pilot to earn money either in spare time or in developing a career in aviation.

You can reduce your costs of flying by taking full advantage of the tax incentives offered by Uncle Sam to encourage the ownership and operation of private and business aircraft. You can use depreciation, investment tax credits, and deductible expenses to lower your tax obligation while you increase your flying budget.

You can study the potential advantages of the leaseback method of writing off much of your flying costs. You can look into the tax savings of leasebacks that offer you an opportunity to operate your own flying business.

That's the aim of this chapter: to show you how the smart aviation consumers reduce their out-of-pocket costs of flying by utilizing both their planes and the tax laws that affect aviation.

Getting Down to Business

About 70 percent of general aviation planes are flown at least partly for business purposes. It may be a weekend pilot who flies to a monthly seminar or a salesperson-pilot who takes clients to lunch by air, or a contractor who flies between job sites in distant towns. In each case the owner is allowed to deduct at least part of the cost of owning and operating a plane as a legitimate business expense.

How can you take advantage of that business incentive? It depends on what you do for a living. Ask yourself:

Does my job require me to travel long distances over which a small aircraft might be more time-efficient?

Can I use my plane or rent a plane to travel to conventions, seminars, regional meetings, or other business-related meetings?

Are there suppliers or customers influential to my business or job who would benefit from a flight in my plane?

Can I use my plane for emergency transportation of people or packages for business purposes?

The best advice for taking advantage of tax laws that govern aviation business expenses will come from your accountant or tax consultant. Invest in a half hour of professional time in which to outline your job and your flying interests. A creative accountant can often find ways in which you can legally and economically combine business and pleasure.

Your Flying RV

You can also lower the costs of flying by planning recreation activities that utilize your wings. Here's how many budget fliers get more for their flying dollars:

Camping by light plane   Hundreds of families are discovering the versatility of small aircraft in choosing and using remote campsites. A pilot skilled in soft-field landing techniques can gently set a bird down in a quiet meadow away from the subdivision campgrounds. Tents that attach to the undersides of high-wing planes are available, and they offer easy access to both the cockpit and the outside world. A small plane can quickly overfly the Friday afternoon commuting campers, and the fliers can be washing dinner dishes in a cool stream before a motorhome finds a level campsite.

Wilderness flying   Other pilots use their wings to land at or near national wilderness areas and then head off with backpacks. If that sounds like your type of flying, remember to respect wilderness airspace. Some areas (noted on sectional maps) won't allow overhead traffic.

Boating by plane   Pilots living in areas with navigable waters make greater use of their wings possible by installing floats and turning their planes into seaplanes. They can cruise or fish from the cockpit of their amphibious aircraft.

Hunting   If you're a hunter, your plane can take you to remote places that are inaccessible by car or pickup. You can also use your wings to lengthen the hunting grounds you can cover in a weekend. You can fly to nearby states in less time than it would take to drive to many nearby hunting areas.

Family recreation   Your wings can keep together family members who are prone to go their separate ways on weekends. The plane can be the focus of family trips that will not only entertain but also instruct in geography, physics, and the natural beauty of your region.

The idea is that greater utilization of your plane means lower per-hour costs. A versatile plane can replace your trailer or motorhome, a second car, a boat or other recreation vehicle and lower the cost of recreation for you and your family. It can also give you an excuse to discover new horizons from a higher vantage point—and to share your preception with others.

Income Opportunities for the Budget Flier

Your plane can help pay for itself with one or more of the many spare-time income opportunities available to the smart aviator. In addition, the part-time ventures can give you more time in the sky to learn and enjoy flying for business and pleasure. Many ventures will pay your way to clouds. Others will even make you a profit above costs and give you a free plane and a second income. All will give you a good excuse to be flying.

Aerial photography   Even with a rented camera and a rented plane you can make enough cash on a weekend to keep you flying for a week. If you're not an expert photographer, a short course at a nearby college and a book such as Frank Kingston Smith's *How to Take Great Photos from Airplanes* will get you started out on the right f-stop. The subjects you can shoot for profit are varied: industrial sites for manufacturing firms or industrial park owners, homes for homeowners, shopping centers, other pilots' planes in flight for their owners, photos for magazines and local advertising agencies, and photos for calendars and picture postcards.

Aviation writing   If you're a competent writer, you can help support your newfound hobby with articles for aviation magazines and newspapers. You won't want to tackle technical articles on aerodynamics yet, but many magazine readers are also new fliers who have the same questions about general aviation that you have. Find the answers from the experts, write about them, and supplement your flying budget. My own flying budget is derived solely from articles and books on general aviation for the consumer such as this one.

Barnstorming   No, I'm not kidding. In fact, none other than Richard Bach, author of *Jonathan Livingston Seagull* and many other fine books, set out for profits and adventure as a barnstormer in the 1960s. Barnstorming can underwrite a flying vacation, especially if you have an older plane in good condition. Read Bach's *Biplane* and *A Gift of Wings* for some of his unique experiences as a modern barnstormer.

Rent-a-plane   You can supplement your income and help pay for your wings by offering your plane for rent to approved pilots. One budget flier I know uses his plane on weekdays and advertises it for rent on weekends at $22 per hour, which is less than many local FBOs charge. All renters must first be checked out, but his plane is usually booked up two to three weekends in advance by regular renters who want to keep the cost of flying from going too high.

Flying billboards   There's also good money in towing or flying advertising billboards. One such billboard, the Skycaster, is slung under the plane from wing to wing and is operated electronically from the cockpit to spell out advertising messages. Simpler billboards are large letters tied together and towed by a plane. In sunbelt states where clear skies offer year-round flying, flying billboards can offer a healthy full-time living.

Instruction   Once you've become an experienced flier with hundreds of hours in the sky, you can start learning how to train others as a certified flight instructor (CFI). Many larger aviation schools offer special courses for pilots who wish to become instructors. In most cases an instructor doesn't earn a big full-time income, but many instructors find the work an enjoyable and profitable part-time enterprise.

Aircraft selling Selling aircraft can be either a part- or full-time enterprise; and if you're a good salesperson and a knowledgeable flier, it can pay you well. People who sell aircraft, like most other salespeople, are paid on a commission basis. One salesman who works weekends only averages an extra $1000 a month while he gets the opportunity to do what he enjoys most: fly a variety of planes.

Flying may excite you enough to consider aviation as a career. Aviation is a major industry that includes not only the many commercial airlines but also business flying, agricultural flying, manufacturing, and development as well as related electronics industries.

As the member of a flight crew, you may be employed as a pilot, copilot, flight engineer, or flight attendant on a major or commuter airline, a pilot or copilot on a cargo airline, or the pilot of an agricultural plane (crop duster).

Many part-time aviators prefer to work on the ground or as part of a support crew. Jobs in this field include many types of mechanics, electronics specialists, hydraulics specialists, and freight handlers.

In aviation administration many careers are available in flight scheduling and related services, marketing, purchasing, and public relations.

In the manufacturing end, thousands of people are employed in both commercial and general-aviation aircraft design, construction, testing, and dealerships.

There are also many career opportunities in the Federal Aviation Administration, which regulates the use of the skies, and the Civil Aeronautics Board, which controls and regulates the commercial airlines in the public interest.

Where can you train for these aviation careers? There are many excellent schools that cater to aviation students. They include those listed in the following table.

Embry-Riddle Aeronautical University
Regional Airport
Daytona Beach, FL 32014

Embry-Riddle Aeronautical University
P.O. Box 2449
Prescott, AZ 86302

Spartan School of Aeronautics
International Airport
8820 East Pine Street
Tulsa, OK 74151

Ross School of Aviation
Riverside Airport
Tulsa, OK 74107

Parks College of Saint Louis University
Cahokia, IL 62206

Northrop University
1108 West Arbor Vitae Street
Inglewood, CA 90306

Sierra Academy of Aeronautics
Oakland International Airport
Oakland, CA 94614

Florida Institute of Technology
P.O. Box 1839
Melbourne, FL 32901

Boardman Flight Training, Inc.
Meacham Field
Fort Worth, TX 76106

Emery School of Aviation
661 Buss Avenue
Greeley, CO 80631

American Fliers, Inc.
P.O. Box 3241
Ardmore, OK 73401

You'll also find many worthwhile aeronautical courses at state and regional colleges and universities. If you're considering a career in aviation, talk your plans over with a college counselor, who can help you decide what education you need and where and how to get it. You can soon be earning a living in aviation and subsidizing your flying budget.

Taxation without Aviation Is Tyranny

Our guide-capitalistic economic system is designed to encourage the movement of products and money through the marketplace. The theory is great; it's the application that sometimes get bogged down. In any case, our legislators have passed laws to encourage the purchase and use of airplanes for business purposes. They have done so by allowing certain tax advantages or tax incentives to those who own and operate aircraft in their businesses.

It's the Internal Revenue Service's job to collect the nation's taxes and watch tax incentives. Internal Revenue Code Section 162 covers deductible expenses related to business entertainment and travel. The code also says that when you use the plane for business purposes you can deduct depreciation and operating expenses and take an investment tax credit on your plane. The IRS further tells you that you must use your wings at least 50 percent for business in order to use your plane as an entertainment deduction.

If you use your wings for business—through your job, a part-time enterprise, or a leaseback—you can take advantage of the tax incentives to lower your flying costs by earning tax credits. The key to supporting entertainment and travel deductions is in keeping good records. Each time you use your plane in your business, you should record:

Amount spent on fuel, landing fees, meals, etc.

Date and time of business use of plane

Place of departure and arrival

Purpose of use

Passengers

Many smart budget fliers keep a special business journal in their log cases for recording business trips as they are made. They also enclose receipts for any business-related purchases over $25. They fully document their legitimate use of their wings for business purposes.

Depreciating Your Wings

The term *depreciation* is used often in business. It means the projected loss of value of an asset due to wear, deterioration, or obsolescence. Real depreciation in an aircraft is not a problem, because many planes last 30 years or more and can often be sold for more than their original cost. However, the IRS says you can depreciate or deduct a percentage of value of your business plane every year. That is, if your plane is purchased at $20,000, used 50 percent for business, and depreciated for five years, you can legally deduct $2000 per year ($20,000 × 0.50 ÷ 5) from your income even though the $2000 never changed hands and probably never will.

There's one snag: recapture. If you depreciate your plane all the way down to zero dollars and then turn around and sell it for $5000, you're going to have to report the $5000 as income and pay taxes on it.

When buying a plane that you will use wholly or in part for business, talk first with your tax adviser to see how you can best take advantage of tax incentives available. Each case is individual.

Earning an Investment Tax Credit

The investment tax credit (ITC) is a little more complex. Oversimplified, the ITC is an incentive for you to purchase a large-ticket item, such as an airplane, for business purposes. For doing so, the IRS will give you a tax credit of as much as 10 percent of the cost of your plane. That is, if you buy a $50,000 plane, you can get a credit of as much a $5000 on your taxes. If your tax bill that year totals $8000, you will have to pay only $3000. The complexity begins in estimating the tax credit and making sure you qualify. Again, a tax adviser can help you with this major tax incentive.

Elementary Deductions

The IRS also says you may deduct certain expenses of operating your plane for business from your income before you figure your taxes. They include:

Fuel and oil
Maintenance
Insurance
Storage
Landing fees

If you use your plane exclusively for business, you may deduct 100 percent of your depreciation and your operating expenses. If you use it partially for business and the remainder for pleasure, you may deduct your business-use percentage of available depreciation and also your fixed costs (storage, insurance, maintenance), as well as all of the operating costs involved in your specific business flights (avgas, oil, landing fees).

Even if you're among the 30 percent of all pilots who never use their wings for business purposes, there are many legitimate expenses you can deduct to lower your tax obligation. They include the following.

Interest paid   You may legally deduct the interest you pay on the financing of your plane as an expense. Have your lender furnish a total of interest paid on your plane at the end of the year so you can claim it on your taxes.

Taxes paid   Not even the federal government wants to charge you taxes on taxes, so it allows you to deduct taxes paid as another legitimate expense and deduction. If you have purchased a plane or parts, you may deduct any sales taxes you've paid. You may also deduct property taxes you pay on either the plane or the land it sits on.

Charitable use   You may deduct the operating expenses of your plane while you are using the plane for charitable purposes such as flying for the Red Cross or as a member of an organized search and rescue team and helping Boy Scouts or Eagles earn aviation merit badges. Get a receipt for charitable work done with your plane and save it for your tax record.

The time it takes to keep good records can earn you tax savings as you fly on a budget for either business or pleasure.

Starting Your Own Flying Business

Many smart budget fliers have lowered their out-of-pocket expenses and taken advantage of Uncle Sam's generous tax incentives to business fliers by starting their own flying businesses: aircraft leasing firms. More commonly the arrangement is called a leaseback, and here's how it works. You purchase an airplane from an FBO and then lease it back on a percentage lease. That is, the FBO will lease or rent the plane to other pilots and pay you a commission on all the money received.

Since the leaseback arrangement is a legitimate business opportunity, you can take full advantage of all of the tax incentives we've talked about: depreciation, investment tax credit, and deductible operating expenses. The actual profit you earn over expenses will be minimal, but the leaseback offers a greater incentive: it can give you a free or very low cost aircraft to fly.

The FBO may, in turn, lease your plane five days a week to a corporation that doesn't want to make the capital investment in a new craft but desires a plane and the fully deductible leasing costs. Your plane is paid for by the corporation's lease payments, and you get to enjoy your craft two days a week for little or no cost depending on how profitable the lease is.

Your FBO may instead use your craft as a rental and book it up at a higher hourly rate. Your profits then depend on how much use your plane gets. The arrangement can be very profitable or expensive. As long as you don't purchase a leaseback plane much more expensive than one you can afford on your own, you will not have to face costs greater than your flying budget.

A third way to go is to lease your plane to a flying club. Depending on the structure of the club, your basic costs will be covered each month by membership dues and the operating expenses will, hopefully, be paid for by member rental fees. By joining the club yourself you may earn enough profits to pay your dues and fees and be able to fly both your plane and the others owned or leased by the club for little or nothing.

You may even decide to do your own renting of your plane through advertising in local newspapers. You can offer your plane "wet" (with fuel and insurance) or "dry" (without) either by the hour or in time blocks. If you decide to pay for your wings that way, you can deduct your advertising costs, depreciation, interest, taxes, and operating costs—even if these exceed your rental income—and also take the investment tax credit. You also have first shot at using your plane and can schedule your flying customers around your own needs. It's an excellent way to make your plane pay for itself.

The Wide Horizon

Many pilots take advantage of the consumer techniques that can lower flying costs—choosing your wings carefully, reducing operating costs, doing some of your own maintenance, and upgrading your wings economically—but it's the smart budget fliers who use their planes to pay for themselves. They utilize their wings more efficiently. They use their craft for business. They develop unique and enjoyable spare-time income opportunities. They consider aviation a career. They take advantage of tax incentives for the smart pilot. They investigate the many advantages of the leaseback system.

Most important, the budget fliers clearly see that the sky has room for the not-so-rich. They learn that flying parallels life in that it can be as simple or as complex as they want to make it. They discover a wider horizon and a new world within as budget fliers.

12 BEST NEW AVIATION BUYS

Choosing a dozen of the best airplanes for the budget flier from today's aviation marketplace is no easy task. More than 140 models from 33 different manufacturers are available.

The first place to start narrowing the field is the price. The cost of a new plane can range from less than $16,000 to more than $2 million. The top line for this guide to economy planes has been drawn at a suggested retail price of under $30,000. Happily, many more than a dozen craft are offered today in the $15,000 to $30,000 base price range.

The next consideration is usefulness to the budget flier. This is more difficult to analyze and apply. Budget fliers are individuals with unique needs and desires. Some will want a basic two-place plane, and others look for a four-seater for business or pleasure. Still other fliers are looking for an aerobatic plane. The dozen best new aviation buys include something from each of those utility categories.

Another major consideration, of course, is economy. The budget flier is searching for a plane that offers both fuel efficiency and economy of maintenance. The planes selected offer between 18 and 32 miles per gallon of gas at cruise speed. The engines are simple, and parts are readily available to cut maintenance costs. All are single-engine planes with fixed-pitch propellers.

To help you choose your individual best new aviation buy, let's see what each of the specifications means:

Price The price given for a plane is the January 1980 manufacturer's suggested retail price for the basic model. The prices are, of course, subject to change, but they will serve for comparison with the prices of other planes in the marketplace. See your dealer.

Engine The two major small-aircraft-engine manufacturers are Lycoming (made by Avco Lycoming) and Continental (built by Teledyne-Continental). They produce the engines for our Best Aviation Buys. The letters at the beginning of the engine designation indicate the configuration or type of engine. The following numbers state the cubic inches of displacement. The last set of numbers and letters identifies the engine accessories, and the final figure is the horsepower rating. As an example, the Bellanca Decathlon offers a "Lycoming AEIO-320-E2B–150 hp," which translates to "Lycoming brand engine with full inverted-flight aerobatic system (AE), fuel-injection (I), and horizontally opposed cylinders with a total displacement of 320 cubic inches and various engine accessories. The engine is rated at 150 horsepower." There are many other engine codes, but these are the only ones used in our Best Aviation Buys planes.

Maximum speed   The maximum speed is the top or never exceed speed under optimum conditions, and it can be used to compare relative top speeds of various planes.

Cruise speed   In most cases, cruise speed is the speed at which the plane will fly when running at the efficient 65 percent of rated power and when trimmed for straight and level flying. Figures may not be exact because of the variation in planes, flying conditions, and manufacturer's testing models, but they are valid for general comparision.

Rate of climb   The rate of climb is the best rate at which the plane will climb measured in feet per minute (fpm). The rate of climb from takeoff varies with density, altitude (elevation and temperature), and airplane weight, so the actual rate for your application may vary. Consult your flight manual.

Takeoff distance   Normal takeoff distances are measured at sea level and are based upon clearing a 50-foot obstacle. Of course, that is a longer distance than would be necessary if no obstacle were present.

Landing distance   The landing distance is the typical number of feet it will take to land the plane at sea level if it must clear a 50-foot obstacle at the approach end of the runway.

Gross weight   The gross weight of the plane is the maximum weight the plane can legally take off with. It includes the weight of the plane, fuel, passengers, and baggage.

Empty weight   The empty weight is the weight of the plane without fuel or useful load.

Useful load   The useful load is the number of pounds of fuel, passengers, and baggage that the plane can lift from the ground and carry safely.

Seating   The plane's seating is the number of seats available in the plane, such as four-place and two-place. Some planes offer optional extra seating such as Cessna 152's kid's seat, but the pilot should never fly more passengers than the plane is designed for. Not only will added passengers change the weight of the plane, they will also change the center of gravity and thus the flying characteristics. An overloaded plane is unsafe. If you need a four-place plane, buy a four-place plane.

Power Loading   The power loading is found by dividing the maximum or gross weight by the maximum horsepower. For the Taylorcraft F-19, as an example, divide 1500 pounds gross weight by 100 horsepower and you get a power loading of 15 pounds per horsepower (15 lb/hp). That figure offers a numeric relation between the loaded weight and how much power is available at sea level.

Fuel capacity   The fuel capacity is the number of gallons of fuel that can be carried in the plane's standard fuel tanks.

Range   The range of a plane is the distance the plane can travel on the available gas supply. There is no reserve amount in a range figure. That is, the number of range miles is the maximum if you run the tanks dry, which you really don't want to do.

Consumption   The rate at which a plane uses fuel is highly variable; it depends on speed, rpm, altitude, and fuel richness. The general fuel consumption figures are offered in gallons per hour (gph) and then in miles per gallon (mpg) at cruise speed. Planes aren't normally rated by mpg; the figure was added so the reader could compare a plane both with other planes and with familiar alternative modes of transportation.

Efficiency factor   Many factors or indexes could be used to rate the relative efficiency of the planes being considered. The simplest would include a ratio of fuel used to fly the plane fully loaded. That is our efficiency factor. It's a relative figure found by multiplying the power loading figure by the fuel consumption, in miles per gallon. The Beechcraft Skipper's efficiency factor, for example, was found by multiplying the power loading of 14.6 lb/hp by the cruising fuel consumption of 18.7 mpg. The total is 273. That can be compared with the Bellanca Scout's 214 and the Cessna 172's 359 in deciding the relative efficiency of the planes. The budget flier must, of course, also consider utility and cost, but the efficiency factor can help you make your decision on the best plane for your needs.

Here are the budget flier's 12 Best New Aviation Buys:

Beechcraft Skipper is a two-place, low-wing, all-metal airplane powered by a Lycoming 115-horsepower engine. It is produced by the Beech Aircraft Corporation of Wichita, Kansas, and it is the only under-$30,000 Beech plane. The Skipper was introduced in the spring of 1979 as a flight trainer and basic aircraft for the budget flier. Beech estimates operating costs, including hangar rental and insurance, at about $20 per hour based on 400-hour-a-year usage. That's a lot of hours for the average budget flier but not for the economy-minded business flier or small flying partnership.

Beechcraft Skipper

Price:	$19,950
Power:	
Engine	Lycoming 0-235-L2C–115 hp
Maximum speed	122 mph
Cruise speed	112 mph
Rate of climb	720 fpm
Takeoff distance	1280 ft
Landing distance	1313 ft
Load:	
Gross weight	1675 lb
Empty weight	1100 lb
Useful load	575 lb
Seating	Two-place
Power loading	14.6 lb/hp
Fuel:	
Capacity	29 gal
Range	425 miles
Consumption	6 gph (18.7 mpg)
Efficiency Factor	273

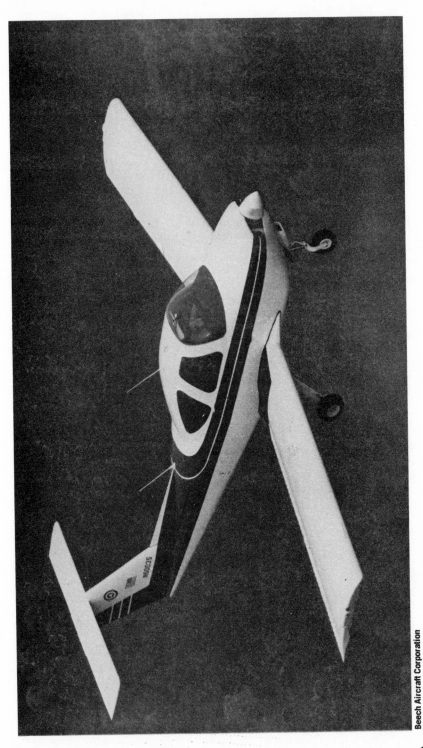

Bellanca Citabria 150 is a two-place, high-wing, Dacron-covered plane with fixed landing gear and a 115-horsepower Lycoming engine. A grandchild of the Aeronca Champion, the Citabria has a large following among budget fliers who fly for fun. It is approved for limited aerobatic maneuvers, yet it offers good gas mileage. The Bellanca name has been recognized among pilots for over 70 years, although the current Bellanca Aircraft Corporation of Alexandria, Minnesota, is, like most aircraft companies, the product of merging many firms and designs. The Citabria carries on the family tradition of low-cost flying. (By the way, Citabria spelled backward is "airbatic.")

Bellanca Citabria 150 (7GCAA)

Price:	$21,950
Power:	
Engine	Lycoming 0-320-A2D–150 hp
Maximum speed	132 mph
Cruise speed	115 mph
Rate of climb	1120 fpm
Takeoff distance	630 ft
Landing distance	775 ft
Load:	
Gross weight	1650 lb
Empty weight	1140 lb
Useful load	510 lb
Seating	Two-place
Power loading	14.3 lb/hp
Fuel:	
Capacity	35 gal
Range	550 miles
Consumption	4.5 gpm (25.5 mpg)
Efficiency Factor	365

Bellanca Scout is a two-place, high-wing, fabric-covered, tail-gear airplane produced by the Bellanca Aircraft Corporation. The Scout is the workhorse of the line. It features a 180-horsepower Lycoming engine, either a fixed-pitch climb prop or a constant-speed prop, oversized tires, and a fast rate of climb. Many Scouts are used as agriculture sprayers or business planes when cargo capacity is needed. As a taildragger the Scout is useful for flying in and out of unimproved fields.

Bellanca Scout (8GCBC)

Price:	$28,500
Power:	
Engine	Lycoming 0-360-C1E–180 hp
Maximum speed	135 mph
Cruise speed	118 mph
Rate of climb	1080 fpm
Takeoff distance	N/a
Landing distance	N/a
Load:	
Gross weight	2150 lb
Empty weight	1315 lb
Useful load	835 lb
Seating	Two-place
Power loading	11.9 lb/hp
Fuel:	
Capacity	35 gal
Range	420 miles
Consumption	6.4 gph (18 mpg)
Efficiency Factor	214

137

Bellanca Decathlon is a two-place, high-wing, fabric-covered, fixed-gear taildragger. It is one of the few unlimited aerobatic competition aircraft produced in this country. It's even approved for inverted flight. The engine is a 150-horsepower Lycoming with a fixed-pitch prop. Options include a 180-horsepower engine and a special aerobatic constant-speed propeller. The Decathlon is nearing its tenth anniversary as a relatively inexpensive new airplane for the budget flier who enjoys snap rolls, outside loops, and other aerobatic maneuvers.

Bellanca Decathlon (8KCAB)

Price:	$26,400
Power:	
Engine	Lycoming AEIO-320-E2b–150 hp
Maximum speed	147 mph
Cruise speed	128 mph
Rate of climb	1000 fpm
Takeoff distance	N/a
Landing distance	N/a
Load:	
Gross weight	1800 lb
Empty weight	1260 lb
Useful load	540 lb
Seating	Two-place
Power loading	10 lb/hp
Fuel:	
Capacity	40 gal
Range	630 miles
Consumption	5.9 gph (21.7 mpg)
Efficiency Factor	217

139

Cessna 152 is a two-place, high-wing, metal-covered, fixed-tricycle-gear plane built by the nation's largest general-aviation aircraft manufacturer, Cessna Aircraft Company of Wichita, Kansas. An updated version of the popular 150 in production since 1958, the 152 is the most popular training airplane in use today. It is simple, economical, and forgiving of the novice pilot. Gas mileage of over 28 miles per gallon at cruise speeds, combined with a high power-loading factor, gives the Cessna 152 the second-highest efficiency factor of our Best New Aviation Buys. An Aerobat model also is available.

Cessna 152

Price:	$16,960 (152-II $21, 300)
Power:	
Engine	Lycoming 0-235-L2C–110 hp
Maximum speed	125 mph
Cruise speed	115 mph
Rate of climb	715 fpm
Takeoff distance	1340 ft
Landing distance	1200 ft
Load:	
Gross weight	1670 lb
Empty weight	1101 lb
Useful load	574 lb
Seating	Two-place
Power loading	15.2 lb/hp
Fuel:	
Capacity	26 gal
Range	480 miles
Consumption	4 gph (28.8 mpg)
Efficiency Factor	438

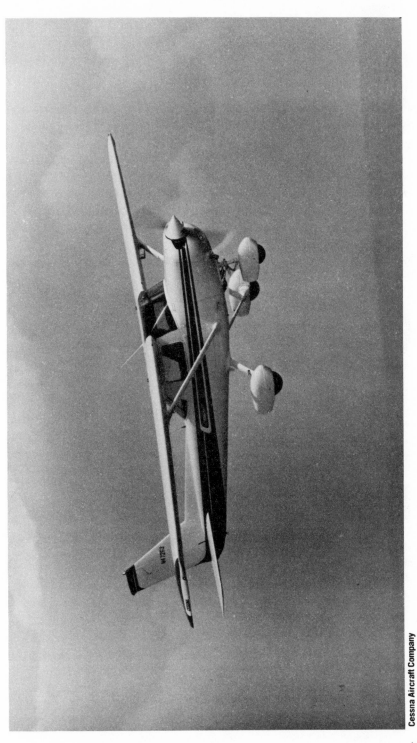

Cessna 172 Skyhawk is a four-place, high-wing, metal-covered, fixed-tricycle-gear aircraft produced by Cessna. In fact, after 25 years in production, it is the most popular airplane on the market; 31,000 units have been produced. The reason seems to be its standing in the middle of the marketplace. It is simple to operate without turbocharging, rectractable landing gears, or other complexities, yet it seats four for family or business flying. Parts and service are readily available. Operating costs are relatively low, and fuel economy is high; it is estimated at nearly 25 miles to the gallon at cruise speeds. The Skyhawk is simply a darn good all-around airplane for the budget flier.

Cessna 172 Skyhawk

Price:	$27,250
Power:	
Engine	Lycoming 0-320-H2AD–160 hp
Maximum speed	144 mph
Cruise speed	137 mph
Rate of climb	770 fpm
Takeoff distance	1440 ft
Landing distance	1250 ft
Load:	
Gross weight	2300 lb
Empty weight	1400 lb
Useful load	900 lb
Seating	Four-place
Power loading	14.4 lb/hp
Fuel:	
Capacity	43 gal
Range	558 miles
Consumption	5.5 gph (24.9 mpg)
Efficiency Factory	359

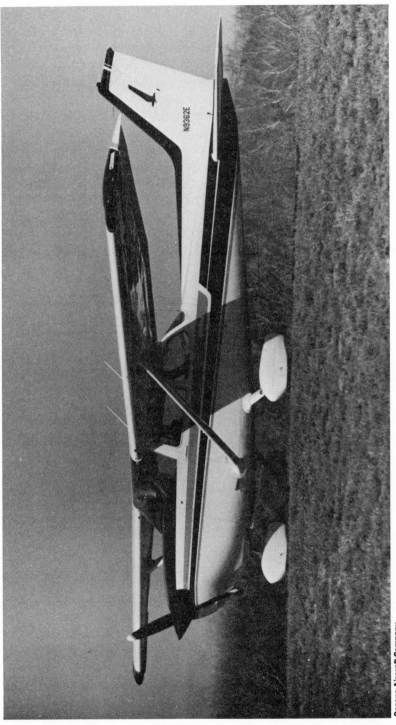

143

Gulfstream American Cheetah is a four-place, low-wing, metal-covered fixed-tricycle-gear airplane produced by the Gulfstream American Corporation—but not for long. Gulfstream, once known as the Grumman American Corporation, is phasing out of the single-engine aircraft market for the larger-aircraft market and is leaving the Cheetah behind. Hopefully, another builder will pick up the popular design. The Cheetah first saw the skies as the BD-1 built by Jim Bede. American Aviation took it over in the mid-1960s as the AA-1 Yankee. American merged with Grumman Aerospace Corporation in 1973, and the Yankee went along for the ride. In 1978, Allen Paulson purchased Grumman American, changed the name to Gulfstream American, and renamed the Yankee the Cheetah. Even as Gulfstream American leaves the Cheetah behind, other firms are considering its continuation. The Cheetah belongs to the budget flier.

Gulfstream American Cheetah

Price:	$30,625 (AA-5A $26,400)
Power:	
Engine	Lycoming 0-320-E2G–150 hp
Maximum speed	156 mph
Cruise speed	148 mph
Rate of climb	450 fpm
Takeoff distance	1600 ft
Landing distance	1100 ft
Load:	
Gross weight	2200 lb
Empty weight	1320 lb
Useful load	880 lb
Seating	Four-place
Power loading	13.3 lb/hp
Fuel:	
Capacity	51 gal
Range	1020 miles
Consumption	5.5 gph (27 mpg)
Efficiency Factor	359

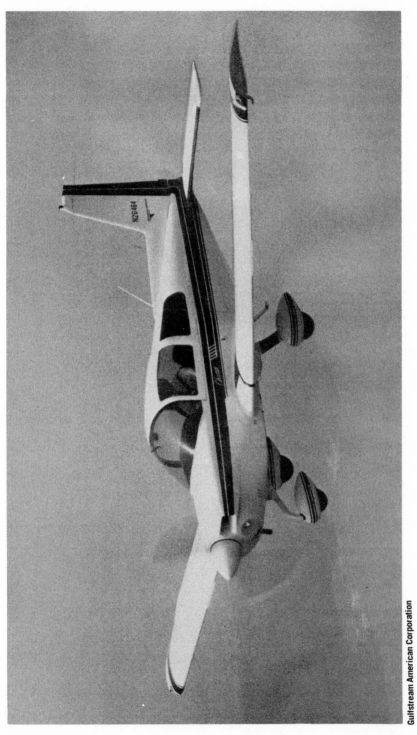

145

Maule M-5 Lunar Rocket is a four-place, high-wing, fabric-covered fixed-gear taildragger with a 180-horsepower engine. Maule Aircraft Corporation of Moultrie, Georgia, claims the M-5 is the only low-priced, truly STOL aircraft in production in the United States today. Its predecessor, the Maule M-4, went into production in 1962 and was replaced by the M-5 Lunar Rocket in 1974. The Lunar Rocket is at the top edge of our price range, but it earns a position in the 12 Best New Aviation Buys of offering seating for four, excellent STOL characteristics, long-range capabilities, a large useful load, and very high cruise and maximum speeds.

Maule M-5 Lunar Rocket

Price:	$29,475
Power:	
Engine	Lycoming 180 hp
Maximum speed	170 mph
Cruise speed	155 mph
Rate of climb	900 fpm
Takeoff distance	800 ft
Landing distance	600 ft
Load:	
Gross weight	2300 lb
Empty weight	1300 lb
Useful load	1000 lb
Seating	Four-place
Power loading	9.8 lb/hp
Fuel:	
Capacity	40 gal
Range	700 miles
Consumption	8.3 gph (18.7 mpg)
Efficiency Factor	183

147

Piper Tomahawk is a two-place, low-wing, metal-covered, fixed-tricycle-gear plane first introduced in late 1977. Designed specifically for flight training and budget fliers, the Tomahawk is an excellent first plane. It is one of the few basic airplanes that isn't directly related to another plane. It is a new design that introduced the GA(W)-1 airfoil and T tail to training planes. The effects are a smaller wing for increased power loading and, many say, better handling characteristics in climbs, descents, and stalls than typical trainer planes can offer. Whatever the reason, the Piper Tomahawk is a very popular plane with budget fliers.

Piper Tomahawk

Price:	$16,840
Power:	
Engine	Lycoming 0-235-L2C–112 hp
Maximum speed	125 mph
Cruise speed	113 mph
Rate of climb	700 fpm
Takeoff distance	1460 ft
Landing distance	1374 ft
Load:	
Gross weight	1670 lb
Empty weight	1080 lb
Useful load	590 lb
Seating	Two-place
Power loading	14.9 lb/hp
Fuel:	
Capacity	32 gal
Range	502 miles
Consumption	4.3 gph (26 mpg)
Efficiency Factor	387

Piper Super Cub is a two-place, high-wing, fabric-covered taildragger produced by the Piper Aircraft Corporation of Lock Haven, Pennsylvania. The Super Cub is simply an updated version of one of the most popular economy planes ever made—the Piper Cub. The Super Cub has been in production since 1949. Originally the power plant was a 90-horsepower engine, but every few years it has been enlarged and now is a 150-horsepower Lycoming. Gas mileage is good; top and cruise speeds are comparable with those of other 150-horsepower engines; power loading is low (primarily because of the useful payload of 805 pounds). The Super Cub is an excellent plane for wilderness and grassfield flying on a budget.

Piper Super Cub

Price:	$24, 520
Power:	
Engine	Lycoming 0-320–150 php
Maximum speed	130 mph
Cruise speed	115 mph
Rate of climb	960 fpm
Takeoff distance	500 ft
Landing distance	725 ft
Load:	
Gross weight	1750 lb
Empty weight	945 lb
Useful load	805 lb
Seating	Two-place
Power loading	11.6 lb/hp
Fuel:	
Capacity	36 gal
Range	460 miles
Consumption	5 gph (23 mpg)
Efficiency Factor	267

151

Piper Warrior II is a four-place, low-wing, metal-covered fixed-tricycle-gear airplane designed to compete with Cessna's Skyhawk. Other than looks, the two planes are very similar: engine, speed, rate of climb, takeoff and landing characteristics, weight and load, and gas consumption. Even the efficiency factors are very similar. The two main differences are the wing configuration and the price. Piper's low-wing version is base-priced over $3200 lower than Cessna's high-wing plane. Both planes are Best Aviation Buys.

Piper Warrior II

Price:	$24,040
Power:	
Engine	Lycoming 0-320-D3G–160 hp
Maximum speed	146 mph
Cruise speed	129 mph
Rate of climb	710 fpm
Takeoff distance	1490 ft
Landing distance	1115 ft
Load:	
Gross weight	2325 lb
Empty weight	1345 lb
Useful load	980 lb
Seating	Four-place
Power loading	14.5 lb/hp
Fuel:	
Capacity	48 gal
Range	810 miles
Consumption	5 gph (26 mpg)
Efficiency Factor	377

Taylorcraft F-19 is a two-place, high-wing, fabric-covered taildragger produced by Taylorcraft Aviation of Alliance, Ohio. The T-craft is the oldest
plane still being built. It all began when C. G. Taylor designed the E-2 and J-
2 planes in the 1930s for the company he owned in partnership with William Piper, the Taylor Aircraft Company. In 1938, Taylor left to start his own
firm and Piper reorganized as the Piper Aircraft Company. Taylor's new firm,
Taylorcraft, began producing a plane very much like the Piper Cub—the
Taylorcraft BC-12D—into the 1950s. In 1974, Charles Feris reopened the
Taylorcraft factory in Alliance and began producing a slightly reworked
plane with a larger power plant, the F-19. The newest Taylorcraft not only
has earned the highest efficiency factor; it is also the lowest-priced of the
dozen Best New Aviation Buys.

Taylorcraft F-19

Price:	$15,750
Power:	
Engine	Continental 0-200A–100 hp
Maximum speed	127 mph
Cruise speed	115 mph
Rate of climb	450 fpm
Take off distance	
Landing distance	
Load:	
Gross weight	1500 lb
Empty weight	910 lb
Useful load	590 lb
Seating	Two-place
Power loading	15 lb/hp
Fuel:	
Capacity	21 gal
Range	500 miles
Consumption	3.6 gph (32 mpg)
Efficiency Factor	480

13 BEST USED AVIATION BUYS

Sometimes a new airplane is out of the question for the flier on a budget; even with financing over 10 years, the cost of a $30,000 plane would be prohibitive. If flying meant a brand new plane, the would-be budget flier would have to take up bowling.

But the would-be flier will be pleased to know that more than two-thirds of the birds flying are more than three years old. The majority of pilots own used airplanes. Why? Because a used airplane makes sense for most budget fliers.

The first year you own it your new plane will cost you dearly in actual depreciation; that is when most planes drop 10 percent or more in value. On a $30,000 set of wings, that amounts to $3000. Depreciation slows down from there, and the plane will actually begin to appreciate in value after eight to ten years. However, most pilots hold their planes an average of less than four years, so the typical pilot loses value on a new plane.

The advantages of owning a new plane over an older one include being able to tailor your plane to your own flying needs and desires rather than those of a multitude of previous owners. You can also take advantage of tax savings in purchasing a new plane that are not available when you buy a used pair of wings. Finally, you get a factory warranty on your plane; and although sometimes there are bugs that must be worked out during the first year, at least the repairs are paid for by the factory.

The used plane offers many advantages, especially to the budget flier. The initial cost and down payment requirement are much lower than those of a new plane. The amount of loss through depreciation is lower; in fact, the value can actually appreciate. Since your plane is worth less than a new plane, your hull insurance is lower than that of a factory-fresh flying machine.

Safety is not necessarily relative to age. Thanks to the FAA requirements for frequent inspections and certified repairs, most planes 20 years old are as airworthy as new planes. More important in considering a used plane is how well the craft was maintained during its life thus far.

The marketplace for used planes, as you discovered in Chapter 4 on buying your wings, is wide open. There are many sources of well-maintained used aircraft for the budget flier who is willing to study the market and shop for the right plane at the best price.

Toward this goal, this chapter offers 12 of the Best Used Aviation Buys in today's marketplace. They were selected from literally hundreds of planes that have tempted fliers over the past 35 years, most of them no longer in production.

The selection criteria for these Best Used Aviation Buys are generally the same as for new aircraft in the preceding chapter: price, utility, and economy of operation. The price ceiling placed on our best used planes is $10,000. Although some approach the top of the scale, many are available in midrange to satisfy even the most economy-minded pilot who doesn't need a complex set of wings to enjoy flying.

Utility was a major factor in selecting many of the planes suggested. The planes range from the simple-fun-in-the-sky Aeronca Champion and Piper Cub to the more modern Beechcraft Musketeer and the top-valued Piper Tri-Pacer. Some are two-place, and some are four-place. Half are taildraggers, and the other half are tricycle-geared. A few are fabric-covered. All are exceptional buys in used aircraft.

Economy is derived partially from simplicity. Many of the planes described in this chapter are basic planes that have seen few modifications in dozens of years of production. Economy in a used plane involves the availability of parts. All of the planes offered here can be rebuilt with parts still in production from various licensees. Some parts are available direct from the manufacturer of the original plane, such as Skyhawk parts from Cessna. Other planes now out of production have national sources for parts still in production. Univair, for example, produces parts for older Aeroncas, Luscombes, Pipers, Taylorcrafts, Ercoupes, Stinsons, and other aircraft. That is important to the budget flier who is considering an older set of wings.

One of the problems inherent in choosing aircraft built over such a wide span of time is the reliability of statistics and performance figures. Once there were fewer industry guidelines for finding, say, the rate of climb or takeoff distance than there are now. The problem is compounded by the many options offered by manufacturers, and even by the number of manufacturers. An example is the Ercoupe produced during the 1940s by ERCO in

65- to 85-horsepower models. Half a decade later the same basic plane was built by Forney Aircraft under the name Aircoupe with a 90-horsepower engine. In 1963, Alon, Inc. began producing the Alon Aircoupe, basically the same plane. A few years later Mooney Aircraft bought the design and modified it into the Mooney Cadet; the Cadet is now out of production and rights to it are owned by another firm. Coming up with reliable performance figures for a plane that has seen more lives than the proverbial cat can be a chore. However, the figures used here are *generally* typical of the subject plane.

Another example is the search for reliable figures on fuel consumption for older planes. Today most manufacturers offer those figures to their customers and dealers, but in 1960 or 1970 the plane builders rightly decided that buyers weren't very interested. After all, gas was inexpensive. Times sure have changed. Many of the fuel consumption figures in this chapter have been reconstructed from manufacturers' fuel capacity and no-reserve range figures. Again, they can be used as a general indicator of fuel consumption and the efficiency factor for the plane.

Here are what I consider as the top 12 Best Used Aviation Buys today:

Aeronca Champion is a two-place, high-wing, fabric-covered, fixed-gear tail-wheel aircraft originally built by the **Aeron**autical **C**orporation of **A**merica from 1945 through 1949. Later, the Champ and other Aeronca planes were built by Champion Aircraft Corporation, which eventually became Bellanca. The early "Airknockers" are still available in great numbers as among the lowest-cost planes to own and operate. If you're not in a hurry and you'd rather fly a simple machine, the Champ is a Best Used Aviation Buy.

Aeronca Champion

Price:	$4500 to $7000
Power:	
Engine	Continental C-85-8F–85 hp
Maximum speed	102 mph
Cruise speed	92 mph
Rate of climb	750 fpm
Takeoff distance	650 ft
Landing distance	875 ft
Load:	
Gross weight	1300 lb
Empty weight	810 lb
Useful load	490 lb
Seating	Two-place
Power loading	15.3 lb/hp
Fuel:	
Capacity	18.5 gal
Range	360 miles
Consumption	19.5 mpg
Efficiency Factor	298

Beechcraft Musketeer is a four-place, low-wing, metal-covered, fixed-tricy-cle-gear airplane built by Beech Aircraft Corporation from 1963 through 1971, although planes of only the first three or four model years are available under $10,000. Powered by a 150-horsepower Lycoming, the Musketeer is a fast little bird with a top speed of 140 miles an hour. The Musketeer is a good buy for the budget flier who has $8000 to $10,000 and wants an economy version of a big name in aviation: Beechcraft.

Beechcraft Musketeer

Price:	$8000 and up
Power:	
Engine	Lycoming 0-320-E2C–150 hp
Maximum speed	140 mph
Cruise speed	123 mph
Rate of climb	740 fpm
Takeoff distance	840 ft
Landing distance	1255 ft
Load:	
Gross weight	2200 lb
Empty weight	1325 lb
Useful load	875 lb
Seating	Four-place
Power loading	14.7 lb/hp
Fuel:	
Capacity	30 gal
Range	354 miles
Consumption	7.8 gph (16 mpg)
Efficiency Factor	235

163

Cessna 120 and 140 are two-place, high-wing taildraggers with metal fuse-lage and fabric-covered wings primarily produced between 1946 and 1951 by Cessna Aircraft of Wichita. The 140 was simply a deluxe version of the 120. Taken together, over 7500 units were produced as simple planes for the expected rush of consumer aviators after World War II. The 140 was later remodeled into a tricycle gear plane and titled the 150. Each generation has earned the respect of thousands of student pilots and economy-minded fliers. (The 140 is shown on the opposite page.)

Cessna 120 and 140

Price:	$5000 to $8000
Power:	
Engine	Continental 85 hp
Maximum speed	119 mph
Cruise speed	103 mph
Rate of climb	650 fpm
Takeoff distance	650 ft (without obstacles)
Landing distance	460 ft (without obstacles)
Load:	
Gross weight	1450 lb
Empty weight	770 lb
Useful load	680 lb
Seating	Two-place
Power loading	17.1 lb/hp
Fuel:	
Capacity	21 gal
Range	440 miles
Consumption	21 mpg
Efficiency Factor	359

Cessna 150 is a two-place, high-wing, metal-covered, tricycle gear plane built for the 19 model years from 1959 to 1977, only to be replaced by a close brother, the 152. There are three versions of the 150: standard, trainer, and aerobat. The standard and trainer are quite similar and vary only in the extra equipment available. The aerobat is a beefed-up version with an airframe that can hold up to most advanced aerobatic maneuvers. The 150 earns a high rating for fuel efficiency and economies of operation that makes it one of the top planes among Best Used Aviation Buys.

Cessna 150

Price:	$4500 to $11,000
Power:	
Engine	Continental 0-200-A–100 hp
Maximum speed	125 mph
Cruise speed	115 mph
Rate of climb	670 fpm
Takeoff distance	1385 ft
Landing distance	1075 ft
Load:	
Gross weight	1600 lb
Empty weight	1100 lb
Useful load	500 lb
Seating	Two-place
Power loading	16 lb/hp
Fuel:	
Capacity	22.5 gal
Range	475 miles
Consumption	5.5 gph (21.1 mpg)
Efficiency Factor	338

Cessna Skyhawk 172 is a four-place, high-wing, metal-covered, fixed-tricy-cle-gear aircraft built since 1956. Only the first six to eight model years will skim under our $10,000 limit for Best Used Aviation Buys, but the popular Skyhawk must be included. They are versatile flying machines that had proved themselves over two and a half decades. All the models in our price range have the Continental 145-horsepower engine, but the main reason for price difference in used Skyhawks is going to be the amount of avionics attached. The budget flier can pick up a good Skyhawk with basic avionics for IFR operation for less than $10,000 and still have room for the family.

Cessna Skyhawk 172

Price:	$8000 and up
Power:	
Engine	Continental 145 hp
Maximum speed	144 mph
Cruise speed	135 mph
Rate of climb	700 fpm
Takeoff distance	1520 ft
Landing distance	1250 ft
Load:	
Gross weight	2300 lb
Empty weight	1400 lb
Useful load	900 lb
Seating	Four-place
Power loading	15.3 lb/hp (150 hp)
Fuel:	
Capacity	42 gal
Range	650 miles
Consumption	15.5 mpg
Efficiency Factor	237

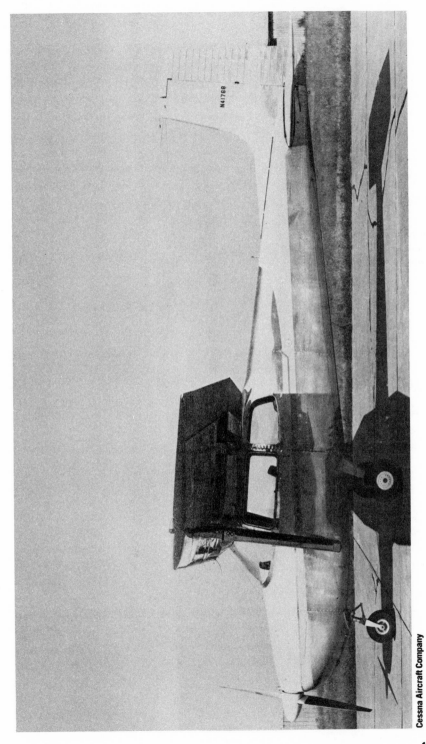

Cessna Aircraft Company

Ercoupe is a two-place, low-wing, metal-covered (early models have fabric-covered wings), fixed-tricycle-gear airplane originally made by the Engineering Research Corporation from 1940 to 1949. The Ercoupe was later produced as the Aircoupe by Forney Aircraft Company and Alon, Inc. More than 4000 Ercoupes/Aircoupes were produced, and most are still in the sky. A simple machine, the Ercoupe was another aircraft designed to help make the transition from driving to flying easy for the American commuter. The controls were interconnected, so the Ercoupe didn't have a rudder control on the floor. It did feature a sliding canopy and other unique features that make it a good investment as well as a Best Used Aviation Buy.

Ercoupe

Price:	$4000 to $6500
Power:	
Engine	Continental 65 to 85 hp
Maximum speed	144 mph
Cruise speed	114 mph
Rate of climb	560 fpm
Takeoff distance	953 ft
Landing distance	1016 ft
Load:	
Gross weight	1400 lb
Empty weight	815 lb
Useful Load	585 lb
Seating	Two-place
Power loading	16.5 to 21.4 lb/hp
Fuel:	
Capacity	24 gal
Range	615
Consumption	25.6 mpg
Efficiency Factor	421 to 525 (depending on hp)

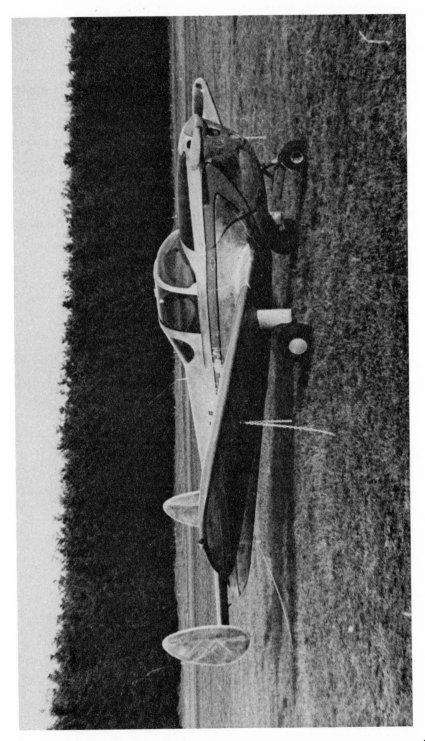

Luscombe Silveraire is a two-place, all-metal, high-wing taildragger built by three different companies between 1946 and 1960. The Luscombe Silveraire (8F) is included among the Best Used Aviation Buys for many reasons. First comes its low cost for a near-classic all-metal airplane. Second, the plane can cruise for 500 miles between fill-ups at 20 miles to the gallon. Third, parts are still readily available through licensee Univair. You can buy a rudder, tail wheel, air filter, or other part from stock. Finally, it has the classic look that many fun fliers desire.

Luscombe Silveraire

Price:	$5000 to $8000
Power:	
Engine	Continental 65 to 90 hp
Maximum speed	128 mph
Cruise speed	117 mph
Rate of climb	900 fpm
Takeoff distance	1900 ft
Landing distance	1540 ft
Load:	
Gross weight	1300 lb
Empty weight	800 lb
Useful load	500 lb
Seating	Two-place
Power loading	14.4 lb/hp
Fuel:	
Capacity	25 gal
Range	500 miles
Consumption	20 mpg
Efficiency Factor	288

Mooney Cadet is a two-place, low-wing, metal-covered, fixed-tricycle-gear plane built by Mooney Aircraft Company after Mooney purchased the assets of Alon, Inc. in 1967. Very similar to the Ercoupes and Aircoupes, the Cadet was an attempt by Mooney to produce a basic plane for the basic flier. Unfortunately, Mooney soon found out that the basic plane cost nearly as much money to manufacture as the more complex plane. After the investment of a great deal of time and money in updating the practical Ercoupe design with a new tail and other features, the Cadet was finally phased out. It's still an excellent buy on the used aircraft market, and it offers the cost-conscious flyer many advantages over the Ercoupe.

Mooney Cadet

Price:	$5000 to $7500
Power:	
Engine	Continental 90 hp
Maximum speed	144 mph
Cruise speed	114 mph
Rate of climb	630 fpm
Takeoff distance	1016 ft
Landing distance	1450 ft
Load:	
Gross weight	1450 lb
Empty weight	950 lb
Useful load	500 lb
Seating	Two-place
Power loading	16.1 lb/hp
Fuel:	
Capacity	24 gal
Range	600 miles
Consumption	25.6 mpg
Efficiency Factor	411

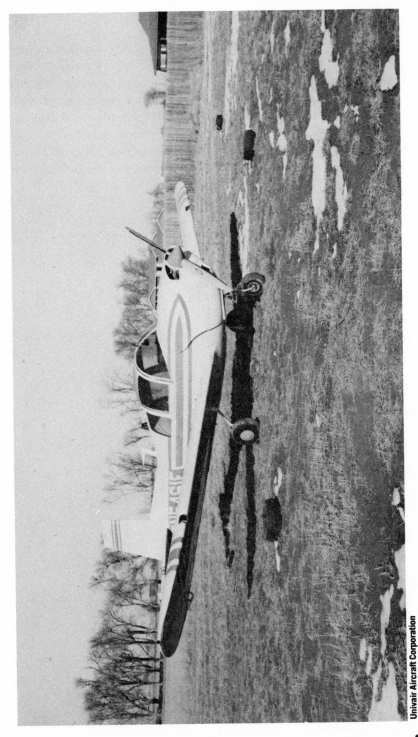

175

Piper Cherokee 140 (PA-28) is a two-place, low-wing, all-metal aircraft with fixed tricycle landing gear which was first produced in 1964 and has undergone many changes since. It began as a 140-horsepower trainer for the first two years of its life and then moved up to 150-horsepower. Names were changed in 1973 to Cruiser and Flite Liner and then, two years later, to Cruiser and Warrior. The 1964 through 1970 models can often be purchased for less than $10,000. In later models an optional twin seat for the rear converts the plane to four-place. An economical and efficient plane for the budget flier.

Piper Cherokee 140

Price:	$7500 and up
Power:	
Engine	Lycoming 140 hp
Maximum speed	144 mph
Cruise speed	134 mph
Rate of climb	820 fpm
Takeoff distance	1700 ft
Landing distance	1080 ft
Load:	
Gross weight	1950 lb
Empty weight	1180 lb
Useful load	770 lb
Seating	Two-place
Power loading	13.9 lb/hp
Fuel:	
Capacity	36 gal
Range	560 miles
Consumption	26 mpg
Efficiency Factor	361

177

Piper Cub is a two-place, high-wing, fabric-covered taildragger built by Piper Aircraft from 1938 to 1947. Its offspring, the Super Cub, is still in production. The Cub is one of the slowest flying machines around—cruising at 75 miles an hour—but its owners don't seem to be in a hurry. Most enjoy the slow, simple movements that made the Cub the most popular trainer aircraft for many years. Parts for the Cub are readily available, and many clubs cater to J-3, J-4, and PA-18 owners. Because of good fuel efficiency and a 44 percent useful load, the Cub offers the highest efficiency factor of any Best Used Aviation Buy.

Piper Cub

Price:	$5500 to $9000
Power:	
Engine	Continental 40 to 65 hp
Maximum speed	87 mph
Cruise speed	75 mph
Rate of climb	450 fpm
Takeoff distance	700 ft (without obstructions)
Landing distance	800 ft (without obstructions)
Load:	
Gross weight	1200 lb
Empty weight	680 lb
Useful load	520 lb
Seating	Two-place
Power loading	22 lb/hp
Fuel:	
Capacity	9 gal
Range	206 miles
Consumption	23 mpg
Efficiency Factor	506

179

Piper Tri-Pacer is a four-place, high-wing, fabric-covered, tricycle gear aircraft built between 1951 and 1960. It is considered one of the best buys in used airplanes today. The efficiency factor is low, primarily because cruising speed is low for the power; consequently, the mileage also is low. To balance that factor, the purchase price is low; and a good Tri-Pacer can usually be purchased for less than a new car. Parts are readily available, and maintenance requirements are usually low. The Tri-Pacer is a popular first plane. Many Tri-Pacers are owned in small partnerships that make an inexpensive four-seat airplane available to budget fliers.

Piper Tri-Pacer (PA-22)

Price:	$5000 to $9000
Power:	
Engine	Lycoming 135 or 150 hp
Maximum speed	140 mph
Cruise speed	130 mph
Rate of climb	750 fpm
Takeoff distance	1600 ft
Landing distance	1280 ft
Load:	
Gross weight	2000 lb
Empty weight	1100 lb
Useful load	900 lb
Seating	Four-place
Power loading	13.3 lb/hp (150 hp)
Fuel:	
Capacity	36 gal
Range	536 miles
Consumption	14.9 mpg
Efficiency Factor	198

Taylorcraft BC-12D is a two-place, high-wing, fabric-covered, tail-gear aircraft built by Taylorcraft Aviation in 1946 and from 1949 to 1954. The T-craft offers a high efficiency factor, excellent fuel consumption, a high useful load for its weight, and simplicity in the sky. Just over 3000 T-crafts were built, and many of them are still flying. The Taylorcraft is again in production, and demand for both new and used T-crafts is high. Prices are appreciating on older T-crafts, which are therefore good investments in low-cost flying for low-time budget fliers.

Taylorcraft BC-12D

Price:	$4000 to $8000
Power:	
Engine	Continental A-65-8–65 hp
Maximum speed	100 mph
Cruise speed	95 mph
Rate of climb	800 fpm
Takeoff distance	N/a
Landing distance	N/a
Load:	
Gross weight	1200 lb
Empty weight	750 lb
Useful load	450 lb
Seating	Two-place
Power loading	18.5 lb/hp
Fuel:	
Capacity	12 gal
Range	300 miles
Consumption	25 mpg
Efficiency Factor	463

INDEX